Save Biodiversity, Save One Billion Children
The Emerald Renaissance Strategy

We need a transformative strategy to save biodiversity in a decade.

The Emerald Renaissance Strategy transforms our shared planet with biosolutions designed to create clean and healthy space for all living creatures while ending waste and pollution.

Mark R. Edwards, PhD

SCAD – Southern Cross Agri-energy Development

Credible Voices for a New Food System

Biodiversity protection is fundamental to achieving health, food security, poverty reduction and more inclusive and equitable development.

Humans have changed ecosystems more rapidly than in any comparable period of time in human history, largely to meet rapidly growing demands for food, fresh water, timber, fiber and fuel. This has resulted in a substantial and largely irreversible loss in the diversity of life on Earth. **– Gary Larson**

If we pollute the air, water and soil that keep us alive and well, and destroy the biodiversity that allows natural systems to function, no amount of money will save us. **– David Suzuki**

We should preserve every scrap of biodiversity as priceless while we learn to use it and come to understand what it means to humanity. **– E. O. Wilson**

The continuing loss of biological diversity will cost the global economy £14 trillion a year by 2050. **– Ecological Institute, EU and US**

So much of the habitat destruction and pollution is based on the simple principle that we somehow have been given free license over other species to degrade the planet. **– Greg Graffin**

Ecosystem services delivered by biodiversity, such as crop pollination, water purification, flood protection and carbon sequestration, are vital to human well-being. Globally, these services are worth an estimated US$125 to 145 trillion per year, more than one and a half times the size of global GDP.[1] **– Secretary-General of the OECD**, 2020

The nation that destroys its soil, destroys itself. **– Franklin D. Roosevelt**

Reducing the Western diet's adverse impact on human health and the environment is one of the greatest challenges facing humanity. **– David Tilman, Regents' Professor, University of Minnesota**

Insect biomass loss has plunged 80% in 30 years. Insect loss threatens the survival of mankind.[2] Unless we change our ways of producing food, insects will go down the path of extinction in a few decades. **– Francisco Sánchez-Bayo** *Biological Conservation*, April 2019.

Food is the strongest lever to optimize human health and environmental sustainability on Earth. We face an immense challenge – to provide our world population with healthy diets from sustainable food systems. The outcome is dire. **– EAT-Lancet Commission**, 2019

Modern food production drives climate instability and ecosystem destruction. A radical transformation of the global food system is urgently needed. Failing immediate action, today's children will inherit a planet that has been severely degraded and where much of the population will increasingly suffer from malnutrition and preventable disease.[3] **— Prof Walter Willet, MD, Harvard – EAT-Lancet Commission**

Industrial food production is a major source of greenhouse-gas, the main user of fresh water, and a leading driver of biodiversity loss and land-use change. Unhealthy foods are behind growing rates of diabetes, heart disease and cancer. We need to change farming and eating if the world is to sustainably feed 10 billion people by 2050.[4] **– The Economist**

No one saves us but ourselves. No one can and no one will. We ourselves must walk the path. **— Buddha**

We are all visitors to this time, this place. We are just passing through. Our purpose here is to observe, to learn, to grow, to love… and then to return home. **— Australian Aboriginal proverb**

The world is one family. **— Vasudhaiva Kutumbakam — Ancient Sanskrit**

Table of Contents

1. Emerald Solutions 7
2. Reverse the Curve 11
3. Biodiversity and Agriculture 13
4. Our Children in Peril 19
5. Fear our Food 27
6. Emerald Forest Initiative 31
7. Emerald ZooPoo Initiative 41
8. Emerald Bioeconomy 49
9. Ecolanda 55
10. Nrich – Superior Nutrition 63
11. Water Energy Nexus 69
12. Ecolanda's Smart Ecocity 73
13. Freedom Foods 77
14. Emerald Flow – Climate Care 83
15. Emerald Ecological Restoration 89
16. Emerald Renaissance 93
17. Human and Environmental Health 101

Mark Edwards, Director SCAD Ecolanda, Director, Biotech and Ecology

Major Contributor: John O'Hare, Managing Director, SCAD Queensland and SCAD Ecolanda

ISBN: 9798729399048

Copyright ©, SCAD Queensland, all rights reserved, 2021.

Save Biodiversity may be used for educational purposes.

The author expresses deep gratitude to the following contributors for technical inputs, data and constructive criticisms which has helped immensely in shaping *Save Biodiversity*.

Acknowledgements

Xuemei Bai	David Punchard	Charles Greene
Robert Henrikson	Zhao Yiyuan	Bruce Rittmann
Stephen Mayfield	Michael Dereel	Ann Ewen
Paul Bryne	John Benemann	Ben Wells
Caroline Hragellund Ruckers	Lieve Laurence	Ira Levine
Anastasia O'Rouke	Maura Flight	Klaus Lackner
Amha Belay	Mimi Hall	Chris Hall
Susan Shultz	Ben Cloud	Efraín Reséndiz
Qiang Hu	Emily McKoy	Cathy Stanton
Antoine Schellinger	Elham Fini	Mona Molnvik
Zuzu Meta	Alice Burns	Andy Ayers
Ras Neillsen	Richard Bellingham	Mark Ewen
Gary Nicholson	Ken Warneke	Rod Missen
Godfrey Dol	Ted Gribble	Lyle Ewanchuk
Gordon Lacey	William Lee	Dato Annie

Forward

Our humanoid species has a problematic history with stealth predators. Through most of our 200,000 years of evolution, stealth predators silently eliminated many of our ancestors.

Stealth predators are silent but deadly. The first indication of their presence may be an instant before death. Humans with weapons have largely eliminated the stealth predation threat from animals by killing over 92% of apex predators.

A new stealth predator has arisen that has killed 1000 times more humans than all animals though human history. Left unchecked, it will kill billions more humans and trillions more flora and fauna.

This silent slayer has invaded and spread clouds of toxic fallout throughout our biosphere. This dirty bomb leaves nowhere to run and nowhere to hide. Fossil agri-chemical fallout kills tiny humans in womb life and leaves millions more seriously disabled throughout their challenged lives. It imposes chronic illnesses on billions of children, adolescents and adults.

Humans created this indiscriminate predator. The Emerald Renaissance Strategy provides a path to eliminate this grave threat to our children, to ourselves and to biodiversity.

Other sources examine the primary causes of biodiversity loss and its terrible impacts. They make policy recommendations to slow the loss. These policies are too slow to save our children.

Save Biodiversity places the primary cause of biodiversity loss on habitat degradation and destruction by our food supply chain. Our food supply itself deserves equal blame for inflicting serious health impacts on our children.

The Emerald Renaissance Strategy proposes that we change our food supply chain from mechanical to biological to save and restore our verdant ecosystems. These actions will save biodiversity and restore their habitat. We also need to change the food itself from the most dangerous in human history to the healthiest.

Some argue that these changes are theoretical, like geoengineering, to reverse climate change. No, they are not. Nearly every action described here has been tested and works. Our task will be integrating these biosystems in an Emerald Circular Economy at scale that we call Ecolanda.

SCAD, Southern Cross Agri-energy Development, led by John O'Hare, designs and serves as the systems integrator for Ecolanda projects.

Save Biodiversity is my 21st book in the *Green Algae Strategy* series that focuses on solutions to world hunger, poverty, jobs, climate chaos and health. Most those books are or will be available free in colour PDF to students, teachers, leaders and those curious at www.Ecolanda.com. We need your passion and your ideas! Communicate your ideas and willingness to engage.

Prior books contained both theory and practice. *Save Biodiversity* comes from 45-years of R&D on 7-generation sustainable systems. While these biosystems have not been tested yet for 140 years, they are real, and they work today.

Forest habitat has been stolen to make way for intensive mechanical agriculture, IMA. Many authors have expressed confidence that innovations such as GMO, precision systems, speed breeding and even organic production can make IMA clean and sustainable. No empirical proof supports these claims. Every eco-trendline displays quite the opposite, especially metrics on the health of our children and environment.

Save and Restore proposes novel biosolutions that have been neither published nor peer reviewed in the current biodiversity literature.

You, the reader, will serve as the first reviewer. Together, we can stop biodiversity loss and save one billion of our children from appalling developmental disabilities, chronic diseases and premature death.

Best green regards,
Mark R. Edwards, PhD, May 2021

1. Emerald Solutions

I have been impressed with the urgency of doing. Knowing is not enough; we must apply. Being willing is not enough; we must do. — **Leonardo da Vinci**

We are not helpless but our children are. Biodiversity loss correlates closely with the loss of our children to developmental disabilities, chronic lifelong illnesses and huge medical costs before premature death. Our children face a gauntlet of chronic health challenges beginning in fetal life, childhood and early development.

Global research shows that 53 million children under 5 suffer from developmental disabilities.[5] One in six, about 15%, of US children have a one or more developmental disabilities. Nearly every medical malady trend line slope rises steeply upwards. Each year more children are inflicted with developmental and medical disabilities that degrade their lives and the lives of their families.

If we fail to act, we will lose one billion of our children to developmental disabilities, chronic life-long diseases and premature death. We will lose them along with animal and plant life.

About 140 million newborns greet our world each year. Every family has high hopes for each precious child. Unfortunately, millions will not be born due to male sperm loss. Additional millions will not survive fetal life or be born prematurely with severe life-long developmental disabilities.

Medical trends indicate one billion of our children are on track for disability and premature death in the next generation from two fixable sources.

1. **Our food supply chain** that is the deadliest in human history for children – and for adults. Never before has food production inflicted so many deadly toxins and contaminants on consumers, ecosystems and biodiversity.

2. **Our food supply**, the food we eat, has become the deadliest in human history. Our food imposes record levels of micronutrient deficiencies, metabolic disorders, obesity, diabetes and other Western diseases.

The food supply chain includes extractive industries that mine and ship agri-chemicals, fertilizer and pesticides as well as petroleum companies that supply electricity and diesel power. The food supply chain delivers inputs to farms, grows feed for animal farms and farms plants and animal crops. The supply chain also harvests, stores, transports, processes, packages, wholesales and retails modern industrial food.

Food supply chain deaths

Intensive mechanical agriculture, IMA, creates a toxic plume that spreads deadly fallout in our air, water and shared environment. No one can escape the toxins that take an especially heavy toll during fetal life. Agri-poisons spread everywhere on wind and water. Contaminants create a fine toxin layer that we ingest when we breathe, drink, touch and eat.

Bioaccumulation amplifies deadly effects. Agri-toxins silently build up in tissues and organs. Too often, the first symptoms display only after the contaminant has caused severe damage and leaves the victim chronically ill or disabled.

Agri-poisons include the CO_2, methane, nitric and sulfuric oxides that drive climate change, create smog and inflict children with asthma. Another deadly vector comes from tiny black soot particulates from combustion in diesel tractors, harvesters and trucks that kill 10.2 million a year and seriously disable several times more people.[6]

Fertilizers are hazardous to our children. Fertilizer runoff causes over 500 dead zones globally. Eutrophication kills all aquatic life over large areas. Dead zones give off massive quantities of CO_2 and methane. These contribute. 20% to total GHG emissions. Incredibly, current climate models ignore agri-generated eutrophication.

Over 5,000 deadly pesticides are used globally. Only about 1% of an applied pesticide is absorbed by the crop. The residual erodes into our shared environment. The majority of modern foods contain pesticide residuals. Most water treatment plants are equipped to remove neither fertilizer nor agri-poisons.

The list of child disabilities and death from our food supply chain is extremely long and heartbreaking. Trends show many of these cases are increasing annually. Our children are helpless to these threats, which include:

- Fetal disability, death or premature birth, life-long developmental disabilities, as well as stunting and wasting.
- Micronutrient deficiencies and metabolic disorders that lead to long-term health afflictions such as obesity, type-2 diabetes, rickets, anemia, goiter, coronary heart disease, cancer, stroke, and osteoporosis.
- Learning disabilities such as ADHD, autism spectrum disorder, dyslexia and disorders with vision, hearing, speech and graphics.

These horrific outcomes often result in children who feel constant fatigue and depression. They have trouble with anger management, family, school and eventually holding a job. Families and society pay an extremely high price supporting the disabled who cannot care for themselves.

The actions we take to save biodiversity are critical to improve human health, vitality and longevity. The actions described here target saving flora and fauna in order to save our children. The actions annually will save 10.2 million human lives, save $2.3 trillion and create over one million sustainable jobs a year.

Taking action to save biodiversity has many benefits beyond saving one billion of our children. These initiatives will:

Save and Restore Biodiversity
- Reverse deforestation in a decade
- Save coral reefs, fin, shellfish and marine life
- End eutrophication and restore aquatic life
- Restore cropland to natural stands

Save and Restore our Environment
- End waste; clean and restore polluted air, water and ecosystems
- Reverse climate change
- Create abundant blue water
- End the use of pesticides and agri-poisons

Good outcomes require the coordinated actions contained in the Emerald Renaissance Strategy.

Emerald initiatives

"Green initiatives" talk of hope to be carbon neutral sometime in the future. Some use carbon offsets from others' actions to make their green pretense. Green strategies fail to address extraction, waste and pollution. Green methods have neither commitment nor methods to save and repair degraded ecosystems or biodiversity.

Emerald initiatives are carbon neutral on day one. They avoid extraction and create zero waste and zero harmful emissions. Emerald biosystems clean industrial pollutants from our air, water and ecosystems.

Emerald biosystems cultivate food, feed, fibre and other bioproducts that are healthier for people, producers, animals and plants.

Growers cultivate freedom foods that do not consume fossil natural resources.[7] Emerald biosystems are healthier for our children and support the repopulation of biodiversity.

Emerald Renaissance Strategy

The Emerald Renaissance strategy includes three layers of actions.

1. **Emerald Initiatives** to halt deforestation, desertification, coral reef loss and the loss of fish and other marine life.

2. **Emerald ZooPoo** to transfer the biosystem knowledge developed in the Emerald Initiatives to farmers, gardeners and students globally through zoos, botanical gardens and science centres.

3. **Ecolanda** to provide global demonstration and R&D sites that implement the Emerald Renaissance to reverse climate change and to save our children along with biodiversity.

Emerald Renaissance

The Emerald Renaissance proposes to transform our food supply from deadly to healthy for people, producers, animals, plants and our shared environment.[8] The Emerald Renaissance replaces mechanical agriculture with emerald biosystems that preserve natural resources with zero waste and negative pollution.

Emerald biosystems produce superior food and other consumer products with microcrops rather than field crops. Each ton of production in emerald biosystems captures, recycles and repurposes two tons of greenhouse gases, GHG.

Ecolanda circular bioeconomy communities are designed, built and operated by **SCAD**, Southern Cross Agri-Energy Development. SCAD serves as the system integrators for these 7-generation sustainable projects. Ecolanda integrates smart agri, energy, waste, water and a smart ecocity that creates zero waste and negative pollution. Negative pollution cleans the environment.

A halt to biodiversity loss, climate change and abundant water begins with Emerald Initiatives.

Emerald Initiatives

The **Emerald Forest Initiative** will help farmers learn to cultivate substitute bioproducts for those responsible for forest and habitat loss – oil palm, soy feed, forest products and plant-based beef. We will cultivate these bioproducts in microfarms at less economic, social and environmental cost than modern mechanical agriculture.

We will create similar Emerald Initiatives to halt desertification, coral reef loss and the loss of fish in the sea as well as marine life. All the products causing those losses can be produced in emerald biosystems that restore rather than pollute ecosystems. These Emerald Initiatives are doable within 10 years.

We will give these novel biotechnologies to farmers so they can grow healthier bioproducts for themselves, family and community. Bioproducts grow without consuming cropland, freshwater, fossil fuels, chemical fertilizers, pesticides or other agri-poisons.

We look forward to "Emerald Soy Day," the day biosystems produce soy at lower cost than commodity prices. This day will echo celebration through our diverse animal and plant kingdoms. We will publicly track global progress toward Emerald Soy, Oil Palm, and Forest Products Days. Plant-based meat prices are already lower than beef, but we need to make PBM better.

Industrial food production decimates our shared environment, biodiversity and our children's brains and bodies. Biosystems clean industrial toxins from our air, water and ecosystems.

Emerald biosystems support our children's health with healthier nutrition and food while restoring biodiversity habitat. Emerald bioproducts avoid using the toxins responsible for children's maladies – diesel, other GHG, inorganic fertilizer, pesticides and other agri-poisons.

Emerald biosystems can create any industrial product with bioproducts that offer superior attributes while they avoid toxic emissions. Emerald biosystems can end habitat loss for flora, fauna and marine life.

Emerald ZooPoo

Tell me and I forget, teach me and I may remember, involve me and I learn.
— **Xun Kuang**, Chinese Philosopher

Experience shows that telling farmers about a better way to cultivate crops creates little change. We will create demonstration biosystems that show and involve farmers in cultivating bioproducts.

Emerald biosystems will be set up in zoos, botanical gardens and other public venues to engage farmers in new cultivation methods. Each ZooPoo facility biocycles waste and recovers, recycles and reuses the carbon and other nutrients. Repurposed nutrients may go into animal feed or organic biofertilizer for plants.

These engagements will emphasize how biosystems reduce economic, health and physical risk for farmers. Farmers will learn how they can create higher, sustainable profits with less physical labor. Farmers will also learn how biosystems benefit their children, family, community and biodiversity.

Ecolanda

Ecolanda provides an opportunity to implement the **Emerald Renaissance,** which transforms our dirty, mechanical and non-sustainable food production system to clean, bio-based system sustainable for over seven generations.

The *Emerald Renaissance* proposes a paradigm shift from fossil IMA to sustainable abundance methods to improve health and vitality for people, animals, plants and our environment.[9] This major transformation may take 20+ years. Smart countries will start now.

We are planning €15 billion Ecolanda projects in four countries. We will use predominately biodiversity-friendly abundance methods that use emerald biosystems. Ecolanda will deliver higher agri-energy productivity than IMA with negative waste. The community will capture 75 million metric tons of CO_2 annually and create 20% more freshwater than the community uses.

Ecolanda integrates agri, energy, waste, water and an ecocity to demonstrate a highly productive circular bioeconomy. Connections across production platforms allow the waste from one to become the nutrients for the next. Systems integration eliminates waste and emissions. Ecolanda will cultivate enough food for five million people. Yet, the most important export will be eco-metrics.

ROSIE

The substantial advantages of biosystems over mechanical systems have not been reported at large scale. ROSIE, Return On Sustainable Innovation in Environment will capture the eco-metrics that allow comparisons between IMA and abundance biological methods.

Ecolanda metrics will drive the transformation from mechanical to biological solutions. Reports will be accessible for research at leading colleges, universities and institutions globally.

Reports will be created by a third-party with access to live on-site monitors and very big data. Industrial Economics, IEc, based in Cambridge Massachusetts has deep experience in ecological assessment, natural resource accounting and sustainability metrics.

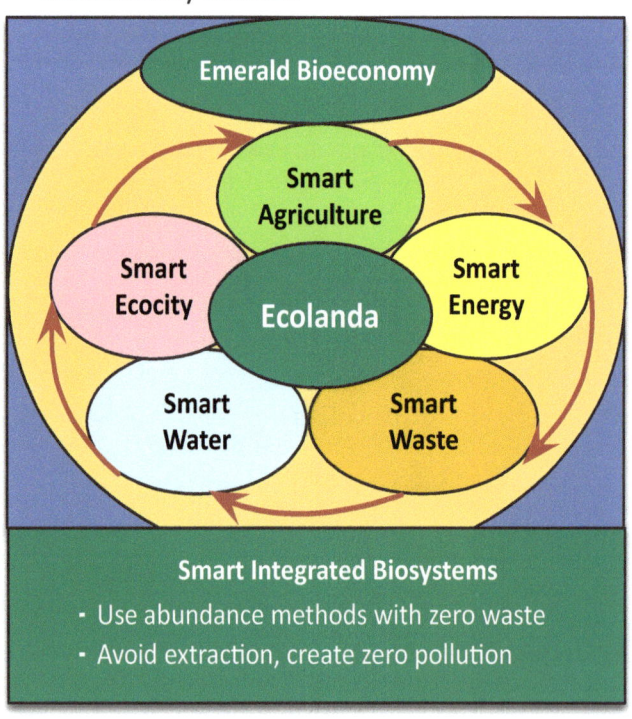

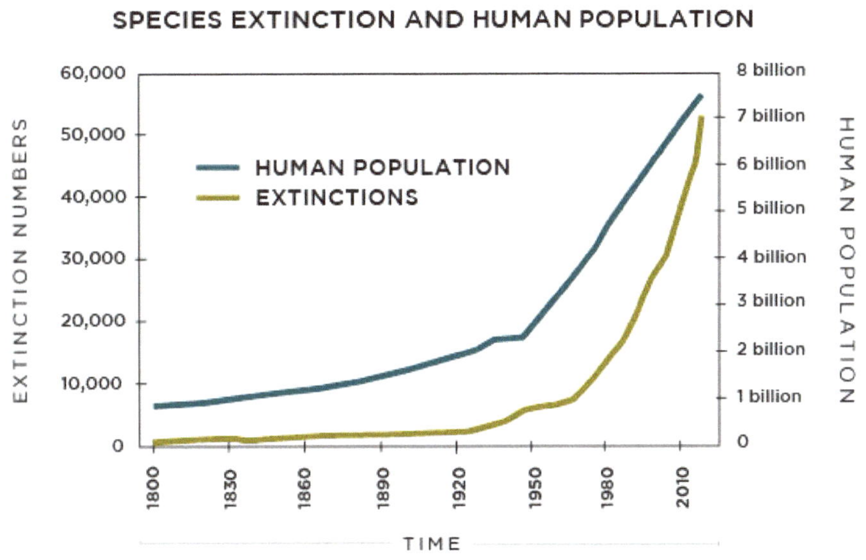

Save Biodiversity

SPECIES EXTINCTION AND HUMAN POPULATION

2. Reverse the Curve

When it is obvious that the goals cannot be reached, don't adjust the goals, adjust the action steps.   **– Confucius**

One million species - out of an estimated total of eight million - are threatened with extinction, many within a few years or decades. We do not have time to "bend" the biodiversity loss curve.

The curves for global warming, GHG, biodiversity loss, environmental pollution and human health are all rising precipitously. We need a strategy to save biodiversity in a decade. Hope in public policies, harvesting limits and education is not a strategy. We need action now in order to save precious biodiversity for our children and future generations.

Reverse curves

The global Living Planet Index shows an average 68% decrease in population sizes of mammals, birds, amphibians, reptiles and fish between 1970 and 2016.[10] Tropical subregions of the Americas suffered a 94% decline, the largest fall observed in any part of the world.

A *Science* study estimates that present rates of extinction are about 1,000 times the likely natural rate of extinction in pre-human times.[11]

The graph shows that extinctions are closely associated with human population growth.[12] The Emerald Renaissance Strategy plans to drop biodiversity loss by 80% in 10 years and to near-zero in 20 years. Our strategy assumes normal human population growth.

State of the biodiversity

Excellent but disheartening descriptions for animal and plant lovers about the sad state of biodiversity loss are available free.

- The Economics of Biodiversity: The Dasgupta Review, 2021.[13]

- Living Planet Report 2020: Bending the Curve of Biodiversity Loss[14]

- Global Assessment Report on Biodiversity and Ecosystem Services, United Nations, May 2019.[15]

- Post-2020 Biodiversity Framework: Targets, Indicators and Measurability Implications, OECD, November 2019.[16]

These comprehensive reports describe the status and the tenuous situation for various forms of life on our tiny planet. They will not be repeated here. Reports recommend "bending the biodiversity loss curve."

What do models current lack?

Those policies are far from sufficient because they lack urgency and the critical actions that will save and restore biodiversity and one billion of our children. Current biodiversity books, reports and policies share similar drawbacks, including:

- **Over dependence on hope** that people will change their behavior. History does not provide positive examples of voluntary behavior change without clear incentives.
- **Operate too slowly**. Biodiversity loss will be too vast before 50-year models have impact.
- **Goals set too low**. Creating carbon-neutral goals are 100%+ too low and slow when carbon-negative options are available. Carbon negative alternatives capture carbon and other nutrients while producing food and other consumer products.
- **Over emphasize GHG**. Greenhouse gasses are less than a 50% solution to clean air. Industrial methods add many other toxins to the air including black soot particulates, $PM_{2.5}$, nitric and sulfur oxides, dust, dander and pesticides.
- **Create tiny change**. Proposing modest social policies around the edges of biodiversity generates meager change. Success favors the bold. Many of the timid in the biodiversity saga are already extinct.
- **Over-expectations for wealthy countries.** Most are unlikely to pay for carbon or biodiversity actions in developing countries. We need actions that countries of all sizes can use to save biodiversity.
- **Fail to motivate**. Asking people to adopt social policies they do not perceive to be in their interest only creates dissonance. People engage when they see actions that profit themselves and their family.

  Peace microfarms in the Emerald Forest section provides an example.[17]
- **Fail to motivate farmers**. Farmers are smart and thoughtful. If making sustainable decisions were easy, they would be using different farming methods and protecting their cropland, communities and biodiversity. Farmers need realistic, tangible action plans to engage in saving and restoring biodiversity.
- **Fail protections**. Creating sanctuaries and harvesting policies sounds great. In practice, reports show they nearly universally lack enforcement and effective protections. Analysis of 44 studies found 40% of 9,000 seafood products from restaurants, markets and fishmongers were mislabeled.[18] Another study calculated that 14 million tonnes of fish are caught illegally every year.[19]
- **Ignore local culture**. Mexico's $3.4 billion Sembrando Vida, or Sowing Life intended to use tree plantings to offset fossil fuel production and use. Direct Mex$4,500-peso (US$230) stipends a month to individuals backfired because historically the community protected the forests. Individuals burned or cut the forest to plant new trees and collect their stipend. Roughly 78 thousand hectares of forest have been lost to date.[20]
- **Fail to engage developing countries**. Countries will only save and restore biodiversity when they see specific actions they can take in their immediate self-interest.

Current biodiversity policy may be compared with dancing with an elephant in the room. Everyone knows that the elephant's behavior creates chaos and destruction. Yet everyone continues to dance along, buoyed by the elephant and the music. We need a biodiversity policy that takes on the elephant.

In the case of biodiversity loss, the elephant is our fossil-based food supply chain known as **intensive mechanical agriculture**, IMA.

Current reports examine biodiversity loss from the perspective of vulnerable animals and plants. Scientists recommend public policies such as sanctuaries, harvest restrictions, monitoring and education to save species. Each public policy offers value, singularly and in combination. They may bend the biodiversity loss curve but just slightly over decades. The animals and plants we love cannot wait that long. Our children cannot wait that long. We cannot wait. We must act!

3. Biodiversity and Agriculture

Health represents the greatest gift we can give and sustain for our children and for our shared environment.

The Emerald Renaissance strategy examines biodiversity loss from a different perspective than other sources – the horrific health risks facing humans, especially our children. Unless we reverse the biodiversity curve quickly, many of us and our children will die alongside our wonderful biodiverse friends.

Plants, animals and humans are systemically losing living space. Space includes physical space to live and the critical natural resources necessary for life – fertile land, clean air, unpolluted water and verdant, unspoiled ecosystems.

IMA waste streams spew massive toxic clouds into our air, water and environment. The meager space left for humans in 2021 has been so severely polluted that no one in the world lives free from pollution. People who live near IMA pollution increase their likelihood of terrible diseases from IMA pollution. Their children are more likely to suffer developmental disabilities, asthma and other illnesses.

Animals and plants on earth evolved based on the availability of clean space, abundant natural resources and clean environments to support life. Dirty IMA systems have changed that.

The Emerald Renaissance biodiversity strategy proposes that we recover and restore space for all life forms and eliminate waste. These actions will provide plenty of space for all living things.

IMA gluttonous consumption

Humans, their crops, and food animals consume half of the world's habitable land, 51 million km^2 (20 million miles^2). About 77% of agricultural land, 40 million km^2 (16 million miles^2) is used for grazing livestock. This massive conversion of forests, wetlands, grasslands and other terrestrial ecosystems has produced a 65% to 90% decline in mammals, birds, reptiles, amphibians and fishes globally since 1970. IMA's fatal flaw is its gluttonous natural resource consumption, below.

Two Agricultural Methods Intensive Mechanical vs Abundance		
Ag method	**Industrial**	**Abundance**
Mechanism	**Mechanical**	**Biological**
Energy source	**Fossil fuel**	**Solar**
All land	50%	0.001%
Cropland	95%	Zero
Freshwater	92%	+ 20%
Fossil fuels	20%	Zero
Inrganic fertilizer	99% 200 MMT	Zero
Pesticides & poisons	99% 25 BKg	Zero

IMA consumes practically all available cropland. Cropland expansion has caused massive deforestation, desertification and biodiversity loss. The need for more food, feed and fibre in coming years will increase IMA resource consumption. A comparison between IMA and Abundance methods displays a stark difference. Abundance methods rely on nature's biological mechanism, photosynthesis, powered by the sun.

A comprehensive study from Cornell University, *The Water Footprint of Humanity,* found that industrial agriculture consumes an incredible 92% of available fresh water.[21] Huge water consumption leaves little remaining for domestic use by humans and far less than needed to support healthy biodiversity. Much of the residual water contains so many toxins, humans and animals risk chronic illnesses from drinking.

IMA uses nearly all the inorganic fertilizer, 200 million metric tons and 25 billion Kg (56B lbs) of pesticides and agri-poisons annually. A majority of these contaminants flow into our environment.

Abundance methods enable food production on non-crop land.[22] Biosystems produce superior food and many other bioproducts while cleaning 20% surplus freshwater. Renewable energy assures no GHG emissions. The only emission is pure oxygen. Nutrients are biocycled, avoiding the use of mined fertilizer. Natural methods are used for pest avoidance, eliminating pesticides.

Widespread adoption of abundance methods will save one billion children in a generation.

Loss of genetic diversity

The UN FAO estimates that 75% of the genetic diversity once found in agricultural crops has been lost over the last century. Just three staple crops – wheat, maize (corn) and rice – now provide more than half of plant-based calories in the human diet.[23] Thousands of traditional crops have been discarded for food production and many have become extinct.

Over 80% of the global food supply uses only 12 crops and five livestock species.[24] Overreliance on only a few species leaves the food system vulnerable to shocks, stresses, pests and creates extreme threats to food security.

Countries that use GMO, genetically modified, crops have cut diversity by 90%. In the US, 90% of maize and 94% of soybean plantings use GMO seeds from a single species. Lack of biodiversity among crops threatens food security, because a single species may be vulnerable to disease, pest vectors, invasive species and climate chaos.

Similar trends occur in **livestock production**, where high-producing cattle, swine, goat and poultry breeds are favored over lower-producing, wilder breeds. Hundreds of cattle breeds have been lost, many with specialized capabilities.

By 2018, the biomass of humans and their livestock (0.16 gigaton) greatly outweighed the biomass of wild mammals (0.007 gigaton) and wild birds (0.002 gigaton).

We have a choice to continue with IMA business as usual or to save biodiversity and our children. We need to transform our food supply to clean, green abundance methods that mimic the sustainable way nature has reliably cultivated food every day for 3.5 billion years.

Mechanical versus biological

The Emerald Renaissance Biodiversity Strategy replaces intensive mechanical with biodiversity friendly abundance agricultural methods.[25]

Mechanical agriculture consumes 10 kilocalories of energy in growing crops for every kcal of food and another 5+ kcal in the food supply chain. Abundance methods use free solar energy with a 1:1 mass balance.

Abundance growers cultivate rootless microcrops like algae, fungi, sea vegetables and yeasts. These crops produce superior nutritious food, oils, feed, fibres and many other consumer bioproducts.

Growers cultivate freedom foods that use zero cropland, freshwater, fossil fuels, chemical fertilizer, pesticides or other agri-poisons. Growers create zero waste as abundance biosystems biocycle carbon and other nutrients from air, water and biological waste streams.

Waste

Waste creates a double cost. Farmers or industrial producers must pay for all the inputs. If like IMA, over 50% of the inputs are wasted, the farmer is forced to pay twice as much as absolutely needed. Farmers must pay again to get rid of animal and biological waste, which is often burned or buried.

Society pays an even dearer price from resource loss, especially when wells run dry. Society pays a terrible price from breathing polluted air, drinking fertilizers and pesticides in well and tap water. Living in toxin-filled environments sentences people to serious disease risk and increased risk for premature death.

No one is spared in 2021 from toxic IMA pollution. People have nowhere to run and nowhere to hide. Modern agriculture produces an appalling pollution bomb that spreads deadly fallout over every part of our shared planet.

Bioaccumulation acts as a stealth predator for fossil food consumers. Their bodies silently accumulate poisons such as lead, mercury and arsenic from air, drinking water and pesticide residuals on and in food. These toxins are ubiquitous in IMA environments.

The redesign of food production to abundance methods eliminates extraction, waste and pollution. Smart, bio-based agriculture safely cleans the fallout left by dirty IMA production. Cleanup takes time because industrial processes have left so much residual waste.

Ecolanda will demonstrate **Emerald Production**, the ability to cultivate superior foods and other bioproducts while cleaning and restoring our environment.[26] When communities understand they have a choice for a cleaner and healthier future, they will no longer accept foul IMA practices.

Every country, region and city create unique priorities for saving what is most important in their community. A good start creates a consensus set of health goals and objectives for people, animals, plants and our environment.

Citizens assess Ecolanda's performance in terms of accomplishing the shared goals.

Emerald Renaissance Biodiversity Strategy

Saving biodiversity begins with 7-generation thinking. Will each system be clean and sustainable for seven generations, 140 years? Thinking 7-generation stewardship may focus on a country, region, city or community.

The man who moves a mountain begins by carrying away small stones. – **Confucius**

Mechanical versus Biological

Intensive Mechanical Fossil foods	Abundance Freedom foods
Mechanical solutions	Biological solutions
Experience = 70 years	3.5 billion years
Large rooted crops	Tiny rootless microcrops
Matures in 120 days	Matures in a day, every day
Mass balance **-10:1**	**1:1**, renewable energy
Waste	**Waste**
Exhaust cropland	Zero cropland
50% of irrigation waster	20% extra freshwater
Massive fossil fuels	Near-zero fossil fuels
60% of inorganic fertilizer	Near zero chemical fertilizer
95% of pesticides, poisons	Zero pesticides, poisons
Pollution	**Pollution**
Dust, black soot, GHG	Cleans air
Emits tons of carbon	Captures tons of carbon
Fertilizer pollution	Cleans air and water
Pesticide pollution	Safely removes pesticides
GMO monoculture staples.	Naturally biodiverse crops
Biodiversity	**Biodiversity**
Eliminates biodiversity	Restores biodiversity
Kills bees and butterflies	Restores insects
Decimates birds and bats	Restores birds and bats
Destroys ecosystems	Repairs our environment
Decimates biodiversity	Restores flora and fauna

The Emerald Renaissance Biodiversity strategy includes the following.

1. **Reimagine renaissance language and art.** Use terms like rebirth, renewal and regeneration that align with 7-generation sustainability.
2. **Rethink priorities**. What human actions cause the most biodiversity loss? What can we do to redirect those actions, so they support clean air, water and environments?
3. **Rue waste.** Eliminate the huge social cost and #1 enemy of biodiversity. Waste is the primary driver of global warming, hunger, poverty and water deficiencies. Circular bioeconomies abolish waste – recovery, recycle and reuse.
4. **Redesign food production**. Intensive mechanical agriculture destroys biodiversity. Transform the massively pollutive fossil-based agriculture with smart, bio-based agriculture that eliminates waste. Smart agri-energy systems cultivate healthier food while cleaning air, water and ecosystems.
5. **Restore health**. What actions enhance health and vitality for people, producers, biodiversity and our planet?
6. **Redirect profit motives**. What actions create profit motives that are win-win for people and biodiversity? How can farmers make profits while saving and restoring biodiversity?
7. **Renew abundant water**. What can we do to restore blue water abundance? What is the most efficient and cost-effective way to produce blue water?
8. **Renovate ecosystems**. What actions go beyond "do no harm" and systemically repair and renew degraded ecosystems?
9. **Restore life**. What are the best ways to repopulate lost biodiversity?
10. **Recognize progress**. Use valid metrics to track and report progress and any gaps remaining.

Each of these actions are explored in more detail in later chapters.

Path forward

Thinking 7-generation stewardship changes the way we plan for resource preservation. Using agri-systems with huge natural resource extraction and waste makes no sense if clean alternatives are available.

Full disclosure. Abundance methods work and have been applied to some degree globally by less than 20 companies. Currently, abundance cultivation costs nearly twice as much as IMA to replace commodities like soy, oil palm and timber. Scale economies are necessary to compete with heavily subsidized IMA commodity products.

Ecolanda will provide the important first demonstration for abundance methods used at scale that can compete and beat subsidized IMA prices.[27] We expect to announce the first economically competitive abundance production of freedom foods within three years.

When abundance methods are competitive on cost, the task will shift to technology transfer globally. We will need to provide the architecture for clean options for farmers and rural communities.

Agricultural methods

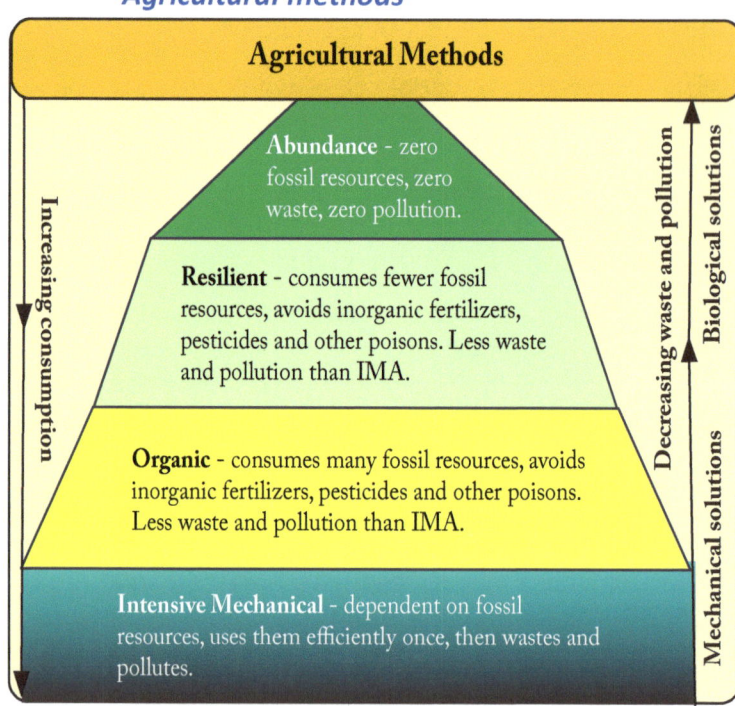

The agricultural methods graphic illustrates the continuum from Intensive Mechanical Agriculture, IMA, to Organic, Resilient and Abundance.

Many people claim organic methods are a solution to IMA waste and pollution. Organic production often delivers 25% less yield. Therefore, it takes more precious cropland from biodiversity. In addition, we would need 10 times more cropland to grow organic compost, which is neither wise nor possible.

Resilient methods targeted to small farmers offer hope for less extraction, waste and pollution.

Controlled environmental agriculture – vertical farms with hydroponics, aeroponics, algae and other microcrop cultivation may practice several elements of abundant agriculture. These farms have green goals for a net-zero carbon footprint.

Biodiversity support by agriculture

Modern industrial agriculture makes the antithesis of positive support for biodiversity. IMA acts as the main driver to destroy biodiversity due to its extremely negative eco-footprint.

Eco-footprint

The eco-footprint metric adds up all the demands a process makes that compete for biologically productive space. An eco-footprint accounts for the amount of the environment necessary to produce the goods and services necessary to support a particular lifestyle.

IMA presents a terrible eco-footprint because the process uses massive amounts of natural resources – land, water, fossil fuels, fertilizer and pesticides – very inefficiently. IMA consumes most the space on habitual land, which leaves insufficient room for healthy flora and fauna.

IMA creates enormous waste and emits vast pollution. Wastes are toxic to humans, animals and plants, which amplify biodiversity loss.

Abundance agricultural methods are the opposite. They mimic nature and use space very efficiently, space that is not crop, forest or wetlands. Abundance cultivates food, feed, medicines and other bioproducts while cleaning pollutants for industrial production. Abundance methods can remediate degraded ecosystems and repair them to enable the return of biodiversity.

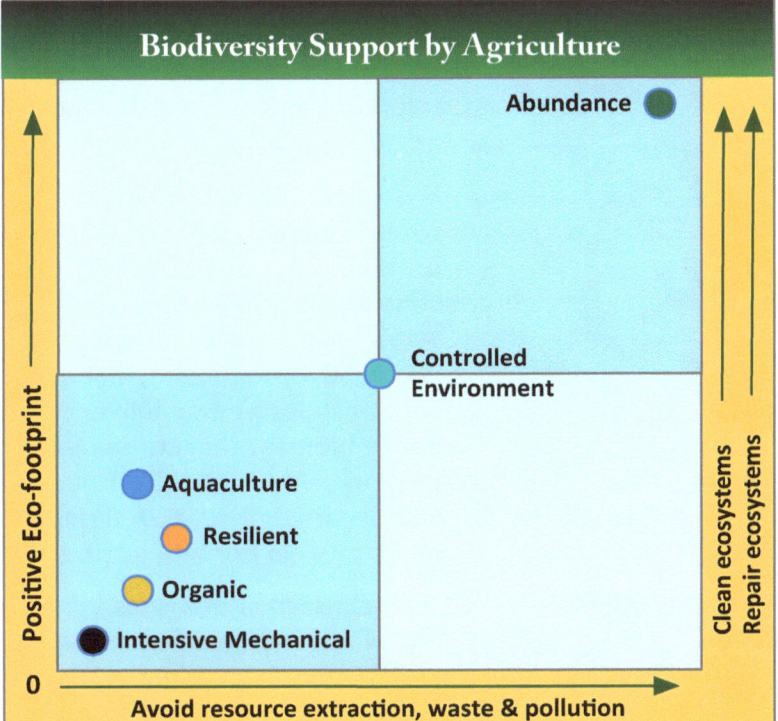

Controlled environment agriculture, CEA, offers a significantly stronger eco-footprint to IMA. CEA uses far less space and creates very little waste.

CEA does not clean and repair ecosystems. CEA will clean and repair the environment in the future as more growers adopt abundance cultivation methods.

Ecological costs

Modern industrial agriculture uses mechanical methods powered by fossil fuels to force food production. Modern farming extracts tons of natural resources and uses them very inefficiently only once. Agri wastes are costly to farmers and to society:

- 50% of irrigation water
- 60% of inorganic fertilizer
- 99% of pesticides

Residuals erode on wind and water and foul air, water and ecosystems. Organic methods moderate some toxic waste, especially inorganic fertilizer and agri-poisons. Ecolanda uses smart agriculture that saves, compared to industrial methods, 75% of water, 90% of inorganic fertilizer and 100% of pesticides.

Controlled environmental agriculture allows growers to consume no cropland and clean rather than pollute air and water. The Ecolanda architecture applies 7-generation biosystems that are living systems powered by free solar and other renewable energy.

The circular linkages assure systems integration which eliminates waste and pollution.

While producing food many times faster than industrial agriculture and field crops like rice, maize and wheat, biosystems clean the air and water. One hectare (10,000 m^2) of algae microcrops can capture 6.7 tons/day of CO_2.

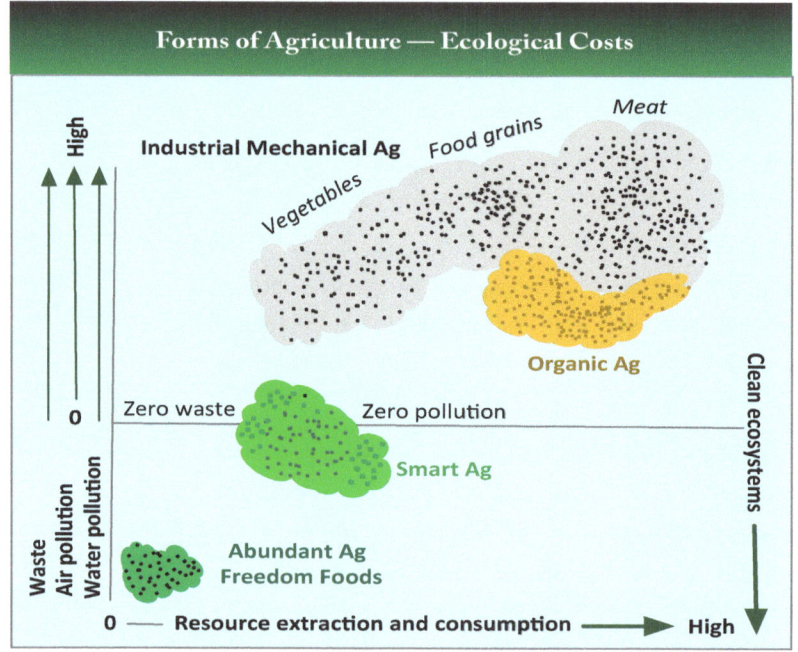

Ecolanda cultivates freedom foods that free growers from the substantial cost of many extracted natural resources and the waste associated with application.

Food grains like rice, maize and wheat deliver only 8 – 11% protein. A kilogram of rice consumes 1,000 liters of fresh water. Some algae species grow 20 to 50 times faster, consume zero fresh water and deliver clean edible biomass that contains 65% protein. Algae-based freedom foods can deliver 6.5 times as much protein per bite as rice.

The next section examines the cost IMA waste and pollution impose on the health of consumers, farmers, animals and ecosystems.

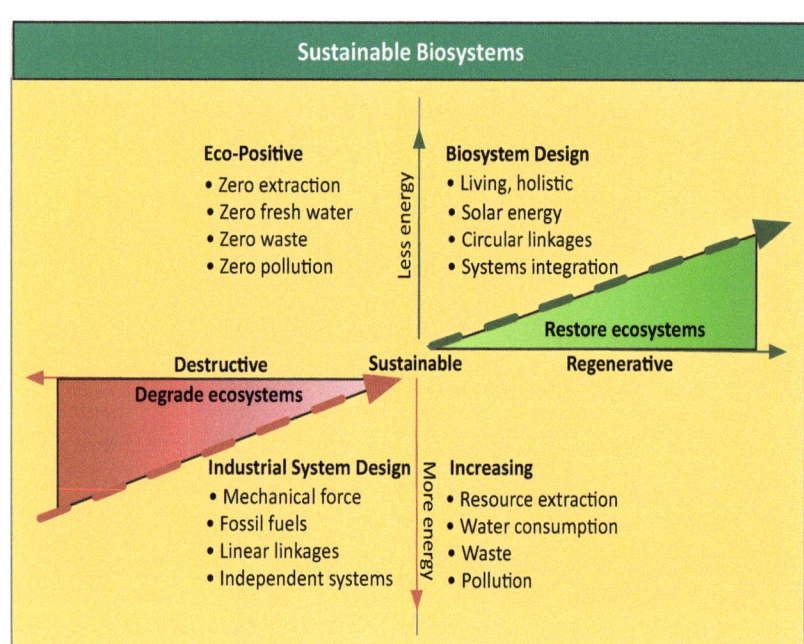

4. Our Children in Peril

We have met the enemy and he is us.
– Pogo

Humans represent only a single of the one million species threatened with extinction. We are particularly vulnerable because we sit at the apex of the food chain. We might think we smarter than other animals. Our collective behavior does not reflect a high level of intelligence. No other animal has done 0.0001% as much to degrade and destroy our biosystem.

In the oceans, 93% of the apex predators are gone.[28] On land, 77% of large animals are in decline and most the residuals are crowded tightly because their historical range has declined by over 70%. Human over-fishing, hunting and habitat loss are the dominant causes large animal loss in terrestrial and marine ecosystem.

Mechanical food production creates a clear and present danger to human health and survival. Humans have corrupted nature's food chain that has reliably sustained life on earth for 3.5 billion years. Intensive mechanical agriculture, IMA, began in 1950 with the transition of the WWII weapons industry to the fertilizer industry.

Fossil Revolution

The "Green Revolution" leveraged synthetic and inorganic fertilizers, NPK, nitrogen, phosphorus and potassium to multiply productivity. Now, 70 year later, many of those high-yield farms and croplands are exhausted and abandoned. Many others are abandoned because water became extinct, or salt invasion or soil compression ruined the cropland.

The agri-revolution was not green. Productivity came from fossil, non-renewable resources. It required an immense expansion in cropland, flood and sprinkler irrigation, fossil fuels, chemical fertilizer, pesticides, and poisons, as well as hybrid and transgenic modified seeds.

The designers of the Green Revolution in 1950 saw starving victims of WWII and decided the solution was calories. Their protein focus created a disastrous error chain that continues today. They selected four **calorie-rich but nutrient-deficient** staples on which to base our food supply – maize, (corn), soy, rice and wheat.

While it is easy to blame the Green Revolution designers for the extraordinary waste and pollution, those were not discussion issues in 1950. Resources were plentiful and cheap and waste was not a problem.

However, over 70 years IMA leaders have only doubled down on while ignoring their obvious mistakes.

Reward system

The reward systems drive behavior, and they surely have for industrial agriculture. Farmers are rewarded based on yield. They emphasize tonnage with no consideration for nutrition. Ignoring nutrition amplified health and environmental problems.

The results have been devastating. Every production decision focused on higher yields, which further diluted non-protein nutrients. Farmers know that adding additional nitrogen causes crops to absorb more water, increasing yields by weight. Consumers pay more for extra water weight but get nutrient dilution. They get fewer nutrients per bite.

GMO decisions centered on pesticide assimilation and yields, and again, disregarded nutrition. Nutrition has been ignored at every step in the development of our food supply.

Mechanical systems

Heavy mechanical tractors, cultivators, trucks and harvesters add dust, black soot particulates and GHG to the atmosphere. Many farmers cross their fields seven times a season, compacting the soil and making it prone to erosion. Machines and fields emit enormous clouds of dust, allergens, particulates and toxic chemicals that create fallout that pollutes everywhere.

Powerful diesel pumps spew black smoke. Pumps pull and flow irrigation water that allowed farmers to substantially expand cropland at the expense of biodiversity habitat. Many communities have seen their wells go dry due to over pumping for crops.

Pumps have also dewatered swamps and wetlands. Wetlands are species-rich habitats that perform valuable ecosystem services – flood protection, water filtration, food chain support and carbon sequestration. Over 50% of worldwide wetlands have been drained for agriculture and their biodiversity lost.[29]

Pests infest

Crop breeding improved yields but created a new problem – more pests. New hybrid and GMO cultivars shifted their energy from pest defense to producing seeds and grains. Both hybrid and GMO crops have weak defenses to insects, worms, weevils and weeds.

Farmers compensate by increasing pesticides, herbicides and fungicides. Agri-poison applications have increased over 500% since 1960. In that time, over 1,000 pest species have developed pesticide tolerance.[30]

Dangerous pesticide residuals remain on 85% of our food supply.[31] Few countries test most foods for pesticide residuals, including the US. Almost no testing has been done on agri-toxin impacts on fetal development and newborns. Current methods for monitoring human exposure to chemicals typically involve screening for only a few hundred of over 8,000 chemicals produced .

Forest loss due to IMA

About half of the world's tropical forests have been cleared, about 1.5 million square miles since 2000. All their stored carbon has been released to the atmosphere. An area the size of Switzerland (38,300 km² or 14,800 miles²) is lost every year.[32]

Forests cover about 30% of global land mass. Three hundred million people live in forests and 1.6 billion depend on them for their livelihoods. Forests contribute nearly 30% of atmospheric oxygen daily while 70% comes from algae, largely phytoplankton in the oceans.[33]

Over 80% of the planet's terrestrial species live in forests. Deforestation creates a major extinction risk to many species.

Save Biodiversity

Plentiful food allowed world population growth from 2 billion in 1950 to 7.9 billion in 2021. Today, IMA faces problems with increasing input costs from nearly all the fossil resources, lack of sustainability, waste and pollution. All four problems impact farmers adversely. IMA waste and pollution are the major cause of biodiversity loss and put human health at extreme risk.

Biosphere contamination

Air, water and ecosystems are shared assets in our public space. IMA effluence flows into our air, water and soil. IMA pollution contributes to an estimated 40% of annual deaths worldwide.[34]

In 2020, levels of heat trapping GHG in the atmosphere reached record high, surpassing 410 parts per million.[35] This trend means that future generations will be confronted with increasingly severe impacts of climate change.

These impacts include more extreme high and low temperature spikes, more powerful storms and floods, long droughts with prolonged water stress, sea level rise and loss of marine and land biodiversity.

The IMA food supply chain contributes 26% to total GHG emissions.[36] Deforestation for agriculture adds 20% to total GHG emissions.[37] Agricultural waste flows do far more damage beyond driving climate change with massive GHG emissions.

Black carbon particulate flows discharged largely from agri-combustion in huge diesel machines are absent from climate models. These tiny particulates do severe harm by accelerating glacier, Greenland and polar ice melt, which amplifies sea level rise.

The tiny black soot particulates rise to the upper atmosphere and travel by winds aloft and drop to white ice sheets. These black fines absorb solar energy and melt the ice like Swiss cheese.

These combustion emissions prematurely kill over two times more people every year than the combined deaths from drugs, HIV/AIDS, smoking, gun violence and traffic fatalities.[38]

Agri-fertilizer flows to waterways. Only about 40% of applied fertilizer gets absorbed into crops. The waste erodes into surface and ground water and causes eutrophication, dead zones, where all aquatic life dies from lack of oxygen. Although dead zones are missing from climate models, a recent study calculated dead zones alone contribute ~ **20% of all GHG emissions**.[39]

Food waste flows through the food supply chain with a 30 to 50% loss. Food waste makes up 28% of municipal solid waste (MSW). Less than 30% of food packing waste is recycled, leaving millions of tons littering our environment and oceans.

Agri-pesticides flow into a poisonous pollution plume. Pesticide fallout is the leading contributor to millions of babies born each year with one or more developmental disabilities. Overspill, irrigation flows from fields, average 46% of all water applied to fields.[40] This adds to agri-poisons that contaminate 80% of our waterways and make many unsafe for human recreation.

Agri-nutrient flows to consumers. The flow of nutrients from croplands has left soils exhausted. Farmers add three macro fertilizers, NPK, but few if any, micronutrients that flow from the field with every harvest. Critical humus, the soil organic material also diminishes. Crops grown in exhausted soil cannot absorb the absent nutrients.

Consequently, over two billion people suffer from life-threatening micronutrient deficiencies due to hidden hunger. Nutrient deficiencies threaten their health, vitality and eventually their life. The nutrient flow in field crops, as well as the dust, dander, mites, allergens and other contaminants, make our food supply unhealthy. IMA food is the most dangerous food in world history. Never before has a food supply carried so many toxins.

Sperm loss

Humans face substantial risk of disability and premature death largely due to toxic IMA waste. IMA over-consumption and waste of precious natural resources causes many serious illnesses. The most significant may be the inability to have children.

Humanity had to endure a deadly coronavirus pandemic in 2021 as well as deaths from climate chaos and toxic air, water and ecosystems. Falling sperm counts also threaten human survival. Toxic agri-chemicals that pervade the space in which we live impede males' ability to produce sperm.

Sperm counts among men in North America, Europe, Australia and New Zealand declined more than 59% between 1973 and 2011. At this rate, half of men would have no sperm by 2045, while many others would have very low counts.[41] Propagation of our species cannot happen if men are spermless.

Reproductive jeopardy

Shanna Swan explains human reproductive jeopardy in *Count Down: How Our Modern World Is Threatening Sperm Counts, Altering Male and Female Reproductive Development, and Imperiling the Future of the Human Race*, (2021). She argues that agri-chemicals are persistent in our living spaces – air, water and in and on food.

Endocrine disrupting pesticides are designed to interfere with insect propagation in order to protect crops. Unfortunately, they do the same for humans and interfere with hormone flows in our bodies. These chemicals essentially block the connection from the brain to the testes.

Failing a signal to produce sperm, men are unable to procreate. The same chemicals disrupt brain, major organ and neurological development in womb-life and often cause fetus death.[42]

Agri-chemicals impose substantial health and disability risks to newborns and young children. Medical research shows exposure associates with attention-deficit/hyperactivity disorder (ADHD). ADHD has increased nearly 50% since 2003.[43]

Children with ADHD suffer severe problems with school, family, employment and community.[44] Adults with ADHD have trouble sustaining jobs, families and relationships.

Endocrine disruptors do their job to interrupt metabolism entirely too well in children. They increase the infliction of obesity, diabetes and other metabolic disorders.

Metabolic disorders, below, increase the likelihood a child will experience severe and repeated illnesses, hospitalization and premature death. Children exposed to endocrine disruptors display a diminished immune response to vaccines, which adds to their health jeopardy.

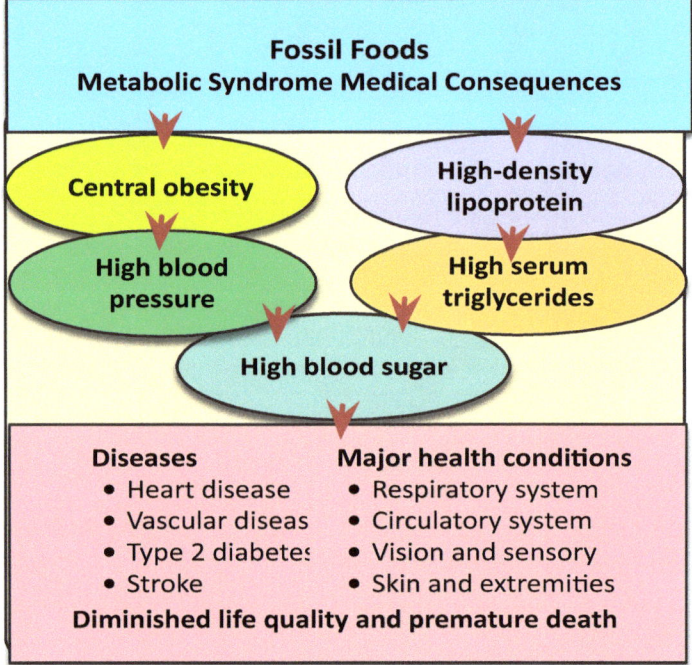

IMA creates massive waste streams that billow into toxic clouds that pollute our shared biosphere. Fertile soils erode on wind and water at the rate of 12 to 20 tons per hectare annually.

A single strong or prolonged storm may erode 120 metric tons of topsoil per hectare. Over 0.4 billion hectares of degraded farmland have been abandoned in the last 30 years.[45]

Erosion often carries away 60% of the farmers' applied fertilizer and 99% of the pesticides.

Agricultural toxic waste contaminates local ecosystems, harms rural communities, and degrades and kills biodiversity. IMA wastes impose a heavy toll on human health, especially during fetal life, newborns, young children and the elderly.

No space to breathe

Exposure to air pollution increases morbidity and mortality for all ages. No humans are free from air pollution exposure. Air pollution causes 17 million deaths worldwide every year and costs an estimated $5.1 trillion in welfare losses.

Fine particulate matter (PM$_{2.5}$)

Fossil fuels combustion in mechanical machines creates airborne fine particulate matter (PM$_{2.5}$) with a diameter of < 2.5 μm. Tiny carbon fines are a key contributor to the global burden of mortality and disease.[46]

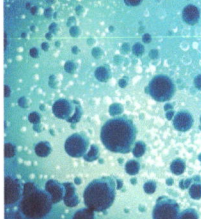

Carbon fines build up in the lungs and body tissues. They accumu-late with other air pollution particles from dust, fertilizer, pesticides and smoke from wildfires and waste.

New research estimates **10.2 million** global premature deaths in 2012 were due to PM$_{2.5}$.[47] Premature deaths occur on average about 10 years before the end-of-life expectancy. The highest mortality occurs in regions with substantial fossil fuel use, China (3.9 million), India (2.5 million), Europe (1.4 million), Asia (8 million), North America (483,000), Africa (194,000) and South America (187,000).

An agriculture transformation from mechanical to biological could save over 100 million life years lost annually from black soot particulates alone.

Developing fetuses and children younger than five years of age are biologically and neurologically more susceptible to the many adverse effects of all air pollutants. Susceptibility to air pollution occurs due to their rapid growth, developing brain, inability to detoxify pollutants, and immature respiratory, thermoregulatory and immune systems.[48]

Children have amplified exposure because they breathe more air per kilogram of body weight than adults.[49] The WHO estimated that in 2012, 170,000 global deaths among children under the age of five were attributable to ambient air pollution. Exposure to PM$_{2.5}$ during pregnancy results in serious language deficits in young children. Language deficits often cause later academic and social adaption problems.[50]

Emissions from IMA outweigh all other human sources of fine-particulate air pollution in the China, India, United States, Europe and Russia.[51] Air pollution from dust, diesel fumes, fertilizer, arial applied pesticides and waste lagoons create toxic clouds of dangerous chemicals.

IMA atmospheric emissions impose severe health risk on rural communities including asthma, wheezing and bronchitis. The closer children live to farms, the more likely they are to experience asthma symptoms. The picture shows an occluded asthmatic bronchiole on the right.

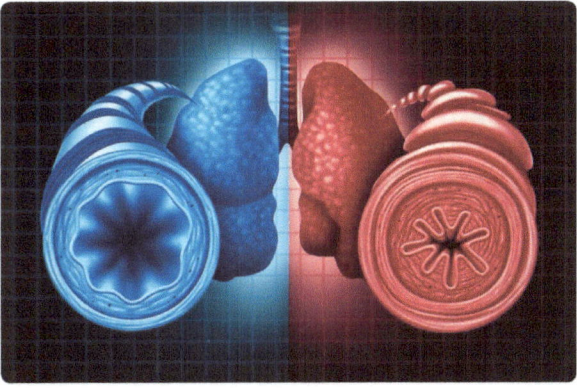

Asthma displays peak incidence in childhood but imposes a lifelong condition for many people. Asthma predisposes children to a myriad of long-term medical impacts including impaired overall health status, compromised social, school and athletic performance and greater disability risk.[52]

A Lancet study reported more than 339 million people suffer from asthma globally.[53] Asthma cases have accelerated with 500 million people expected to be asthmatics by 2025. Developed countries with more IMA have higher rates of asthma for all ages, especially for children. Asthma adds an average burden of $3,100 a year for a person treating the disease.

Livestock

Half of the world's habitable land is used for agriculture. Farmers use most of this land to raise meat and dairy animals. Livestock are the major driver of deforestation and biodiversity loss.

Livestock are fed from two sources, grazing land and field crops, such as soy, maize and wheat. Livestock consume so much land because it takes over 100 times as much land to produce a kilocalorie of beef or lamb versus plant-based alternatives.

Manure emits extensive amounts of nitric oxides and ammonia. A cow burps and farts daily 6,137 liters of CO_2, 5,279 liters of CH_4 (methane) and lots of ammonia, depending on their diet. A cow's daily gas would fill 11,400 one-liter soda bottles.

Livestock account for between 14.5% to 50% of global GHG emissions, depending on the GHG accounting method. Livestock produce 9% of the CO_2 but a much larger share of even more human-induced and harmful gasses, including:

- 65% of nitrous oxides which create smog and have 300 CO_2e (carbon dioxide equivalents)
- 37% of methane which creates smog and pollutes with 39 CO_2e (CO_2 equivalents)
- 64% of ammonia, which gives off a stench and creates acid rain

Livestock urine and manure emit over 400 different harmful gasses into the atmosphere, including nitrous oxide, ammonia, particulates, endotoxins, and hydrogen sulfide.[54]

People who live near industrial farms have a much greater risk from respiratory diseases, bronchitis, chronic asthma, toxic dust syndrome, lung inflammation, immune suppression and mood disorders. They are at higher risk for stroke and cardiac arrest. Ammonia emissions cause dizziness, nausea, eye irritation and respiratory illness. Hydrogen sulfide may cause sore throat, seizures, comas and premature death.[55]

Pregnant mothers who live near farms have an elevated risk for fetal death, premature birth and newborns born with brain, sight and chronic neurological disorders.

Agri-poisons

Epidemiologic studies link pre- and postnatal exposures to pesticides to fetal death, preterm birth, low birth weight and size, intrauterine growth restriction, severe birth defects and developmental disabilities.[56]

Pesticide exposure from air, water, food or households impose adverse health effects on fetal growth, neurodevelopment, respiratory and immune health and childhood growth and obesity.[57]

Children exposed to pesticides either in utero or during other critical periods may have lower IQs, birth defects and developmental delays. They suffer higher risk of autism spectrum disorder, ADHD, breathing diseases and cancer.[58]

Meta-analyses show adverse health effects from even low exposure levels to polychlorinated biphenyls (PCBs) and the neurotoxic effects of lead, methylmercury, PCBs and organo-phosphate pesticides.[59]

No space to drink

> *Thousands have lived without love, not one without water.* W. H. Auden

Farms discharge millions of tons of drug residues, agrichemicals, fecal matter, sediments and saline drainage into waterways. Polluted surface and well water caused 2 million deaths in 2015. Every year, unsafe water sickens over one billion people.[60] Unsafe water kills more people yearly than war and all other violence combined.

The FAO *More people, more food, worse water? A global review of water pollution from agriculture* notes that IMA causes 80% of water pollution in many regions.[61]

Water related Illnesses and access to clean toilets are responsible for the deaths of over 800,000 women around the world each year.[62] The UN *Sick Water* report found that nearly 4% of all deaths are attributed to water-related diseases. This translates to millions of deaths a year. More than half of the world's hospital beds are filled by people suffering from water-related illnesses.

More than 80% of the world's wastewater flows back into the environment without being treated or reused. In some developing countries, the untreated flowback exceeds 95%.[63]

Polluted water transmits diarrhea, cholera, dysentery, typhoid and polio. Contaminated drinking water cause 830,000 deaths from diarrhea each year.[64]

Irrigated cropland has more than doubled in recent decades, from 139 million hectares in 1961 to 320 million in 2012. Irrigation transfers IMA pollutants to surface and well water.

Livestock has risen from 7.3 billion units in 1970 to 32 billion. Livestock production consumes for 70% of all agricultural land and 30% of the planet's land surface.[65] The environmental footprint of meat and dairy is high mainly due to the feed required by livestock in combination with the impacts of cultivating, transporting and processing animal feed. Soy accounts for most the animal feed for animal products like beef, chicken, pork, salmon, cheese, milk and eggs.

A single cow-calf pair gives off more CO_2e methane than a car emits CO_2 in a year traveling 30,000 miles. The overuse of antibiotics in animal production has accelerated the development of antibiotic-resistant bacteria, which cause 76 million cases of food-born illnesses a year and over 5,000 deaths.[66]

Meat-based diets are linked to 60% of global biodiversity loss.[67] The many calls for people to change to plant-based diets have largely failed.

In the future, some communities will require consumers to pay the full eco-footprint cost of meat, which will change consumer preference for animal meat. The other breakthrough, plant-based meat is making excellent progress today.

Dangerous chemicals in newborns

Current methods for monitoring human exposure to chemicals typically screen for only a few hundred of the 8,000 chemicals produced every year due to instrument limitations.

A 2021 study using a new chemical detection instrument discovered over 50 new chemicals lurking in the bodies of newborns.[68] Most had never been observed before in people.

Little is known about these chemicals and their impacts on the human body, especially newborns. Some of these chemicals are traceable to processed foods, agrichemicals and pesticides.

Perfluoroalkyl and polyfluoroalkyl, PFAS, were found in over 1,000 foods.[69] Animal studies show PFAS can cause damage to the liver, immune and cardio systems and cause cancer. PFAS cause birth defects, low birth weight, delayed development and fetal deaths in laboratory animals. Phthalates are present and are known to leach from fast-food boxes. People who eat fast-foods are 24 to 40% more likely to have high levels of phthalates. Phthalates are linked with reproductive problems and lower IQ in children.[70]

Improvements in chemical detection instruments will create an improvement in understanding on how deadly agri-chemicals enter and accumulate in the body tissues during fetal and later life.

Dead zones

IMA practices create over 500 dead zones globally which are expanding about 10% each decade.[71] They emit massive levels of CO_2 and methane as marine life dies and rots The Gulf of Mexico dead zone covers 7,000 square miles.

Major extinction events in Earth's history have been associated with warm climates and oxygen-deficient oceans.[72] We are experiencing both, significant warming and expansion of dead zones.

Dead zones like the Gulf of Mexico, above, continue to enlarge rapidly due to IMA farming

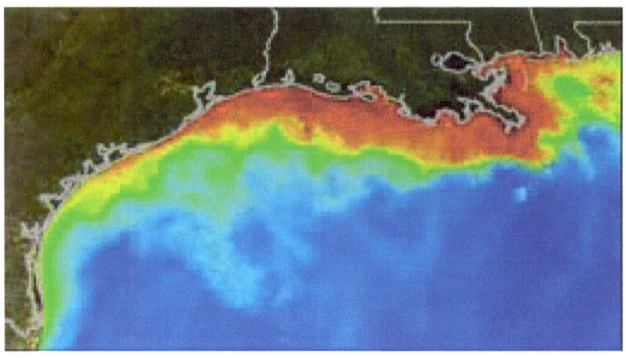

practice. Farmers exhaust the nutrient loads and humus in their fields with continuous production of a single crop rather than practicing crop rotation. Worn out soil requires continuous increases in both fertilizer and water to produce the same yield. A majority of the excess fertilizer runs off into waterways. This adds to the environmental toxin load and expands dead zones that kill all aquatic life.

Water scarcity

The gap between water supplies and water demand is increasing in over half the world. Increasing heat and drought escalate the water gap and act as the major constraint to IMA's ability to deliver sufficient good food.

Agriculture creates severe water scarcity. IMA uses 92% of the available fresh water for irrigation and their animals.[73] Many people must spend hours daily searching for and hauling scarce water.

The amount of water used to irrigate one hectare of maize in a hot growing region, 12,3 million liters, would support 800 urban families. Nearly half of the world's population lives in water-stressed areas. Yet, American farmers use over 3 trillion gallons of water (11.4 trillion liters) annually to grow corn for biofuel production.[74]

The primary IMA innovations over the last 30 years have focused on higher productivity. Hybrids, GMO and more efficient use of inputs typically intensify plantings, which requires more irrigation more often because the plants are putting their energy into growing more grain, not deeper roots.

Farmers tend to waste water because they often pay less than 5% of domestic water rates. About 10% of IMA water evaporates in reservoirs, lakes and canals. Farmers waste over 50% of applied water to evaporation with surface or sprinkler irrigation. Additional water percolates below the root zone where it is unusable by the crop.

Water is bulky and heavy, one kilogram per liter. Storing and moving water for irrigation costs the public billions. Farmers pay only a tiny fraction of the total water cost. The huge pumps that move water and the electricity to power the pumps add substantial GHG to our atmosphere. These pumps are absent in current climate models.

Cape Town South Africa nearly ran out of water in 2018. Other major cities in jeopardy of running out of water include Mexico City, Tokyo, São Paulo and Melbourne, Australia.

Food supply chain summary

Food production using IMA methods puts humans and especially our children in extreme peril. Industrial food production and processing create a plume of toxic fallout across our ecosystems that inflict children and people of all ages with chronic, debilitating health conditions.

These life-long afflictions destroy children's ability to meet their true destiny and increase health costs.

IMA has destroyed biodiversity habitat and poisoned billions of animals and plants. We need a biodiversity friendly method for growing food.

Only about 50% of our children's health risk comes from the food supply chain. Many smart scientists believe the food itself puts children in higher peril than IMA's extreme environmental pollution.

The next section examines our food and its impact on health.

5. Fear our Food

An international team of over 30 medical and nutritional scientists examined all available dietary nutritional and medical research and concluded in *EAT-Lancet: Food, Planet, Health*:

> Unhealthy foods *now pose a greater risk to morbidity and mortality than* **unsafe sex, alcohol, drug and tobacco use** *combined.*[75]

The designers of the Green Revolution made a grave mistake. They bet the new food system on calories but ignored nutrition. They chose calorie dense but nutrition-light food grains like maize, soy, wheat and rice. These fossil foods are grown with mechanical means powered by fossil fuels.

Every food industry decision since 1950 has focused on how to produce more yield. These decisions usually came at the expense of nutrition. Each productivity improvement increased mass but diminished nutrient density. The additional mass that helped farmers gain more revenue was water and non-digestible cellulose. Both compounds diluted the already sparse nutrients in food grains.

Scientists proved over thirty years ago that protein was only one of over 70 nutrients the body needs to sustain health and vitality. Yet, IMA stayed focused on protein and yield.

Their focus on weight has led to devastating penalties, especially for our children. Nutrition has been ignored at every step in the development of our food supply.

Over 70 years later, our calorie-dense but nutrient deficient fossil foods diet has catastrophically damaged our health and the future of our children. IMA produces dirty fossil foods using fossil fuels and other fossil resources.

More people are malnourished today, over 3 billion, than at any time in world history.[76] Nearly one billion are severely undernourished. Globally, one-in-four people suffer from moderate or severe food insecurity.

Approximately 33% of children under age 5, 220 million, are malnourished – stunted, wasted or overweight due to poor nutrition. Two thirds, 440 million, children are at risk of malnutrition and hidden hunger because of the poor quality of their diets.[77] The impacts of stunting on child development are largely irreversible beyond the first 1000 days of a child's life. Stunting has severe impacts on both cognitive and physical development throughout an individual's life.[78]

Fossil foods

Over 60% of the calories consumed today come from only three low-nutrient foods; wheat, rice, and maize (corn).[79] These foods are too hard, starchy and bitter to eat raw. Industrial food processors mechanically strip out the healthy fibre and nutrients and replace them with salt, sugar, thickeners, artificial flavors and additives.

These additives make junk foods attractive and even addictive to many consumers. Uniformed consumer judge food by taste. Poor food labelling provides little information on nutrition or health.

Over 20% of deaths worldwide are attributable to the unhealthy Western diet of fossil foods – high in calories, carbohydrates, sugar and salt but nutrient deficient.[80] The CDC predicts that 1 in 3 children born after 2000 in the US will become overweight and diabetic, 45 million children.[81]

Fossil foods have created a global obesity epidemic where 2.2 billion people are overweight. This is a 300% increase since 1975. Unhealthy foods have driven a tenfold increase in obese children.[82] Obesity and overweight rates for American adults also doubled to 72%.[83]

American children suffer from one of the highest overweight rates worldwide, 33%. The obesity prevalence among US children is 200% higher than among Western Europeans.[84]

Obese people have increased risk for all-causes of death, hypertension, hyper lipidemia, type 2 diabetes, coronary heart disease, stroke, gallbladder disease, arthritis and respiratory illnesses. Obesity lowers life quality, increases body pain and mental illness such as depression, anxiety and other mental disorders.[85]

Fossil foods have decreased the cost of food in the US to about 9% of the household budget, but the high-caloric, nutrient deficient foods have doubled healthcare costs to above 18%.[86] Junk food comprises 86% of ad spending on minority targeted television.[87]

Unsurprisingly, minorities suffer higher obesity and diabetes rates than do non-minorities.

Half of US adults over 50 suffer from a chronic health condition. Over 1 in 4 suffer from two or more chronic conditions.[88]

Hidden hunger

Why are children and family members not getting good nutrition? The root cause is hidden hunger from nutrient deficiencies. Fossil foods suffer from hidden hunger, which results in foods with empty calories.[89] The term "hidden" refers to produce that shows few visible nutrient deficiency symptoms. IMA staple crops commonly lack of one or more key nutrient, particularly iron, magnesium and zinc, as well as essential vitamins, especially Vitamin A. Hidden hunger inflicts severe disorders on consumers.[90]

The nutrient content may be above the deficiency symptom zone needed for passable appearance, but below the zone for optimal crop health. IMA farmers grow the same crop year-after-year, to maximize profits. Predictably, systemic extraction removes vital soil nutrients.

Farmers are paid by weight, not nutrition. Consequently, many fossil foods are nutrient deficient because farmers hold back fertilizer, especially micronutrients, to save money. Most farmers do not replace micronutrients to fields because those formulations may not be available. Applying only macro fertilizers, NPK, may represent 30 to 40% of the cost of the crop. The

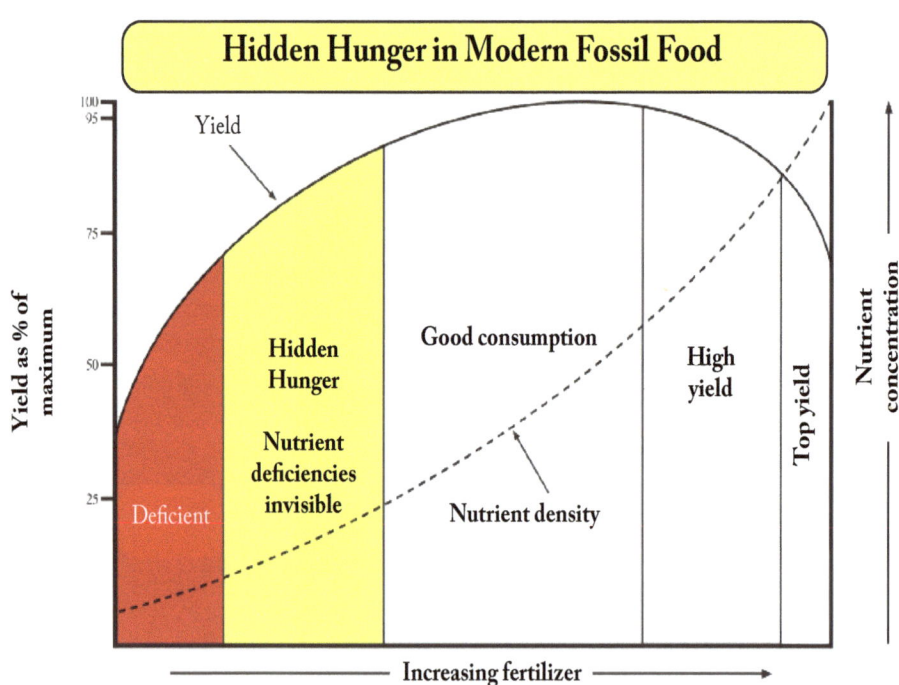

avoidance of addition of micronutrients to fields makes sense from a farmer's cost perspective.

Vegetables, grains and fruits prefer but do not need, a full complement of micronutrients for growth and development. When micronutrients, minerals, vitamins and trace elements have been extracted from years of growing crops without replacement, produce exhibits hidden hunger.

Produce appearance can be deceiving. Field tomatoes look puffed up (with water) today compared with 30 years ago, but the taste, texture and nutrition has degraded by 40%.[91]

Hidden hunger in foods occurs when farmers grow field crops in exhausted soils. Crops grow in degraded soil, but the food lacks micronutrients. Micronutrient deficient produce pass on their deficiencies to human and animal consumers.

Most people cannot taste or feel any immediate affects from hidden hunger. However, fossil foods deficient in micronutrients have substantial impacts on health and create a condition known as micronutrient deficiencies, MND.

Micronutrient deficiencies

Micronutrients are vitamins and minerals that are vital to healthy development, disease prevention and wellbeing. With the exception of vitamin D, micronutrients are not produced in the body and must be derived from the diet.

Micronutrient deficiencies occur in humans from two sources, hidden hunger in field crop produce and mechanical food processing.

Humans evolved eating healthy whole foods such as fruits, vegetables, meat, nuts, honey, beans and seeds. These foods provide dietary fibre and a diverse set of essential nutrients. Processed industrial foods have removed the fibre and many nutrients to make the food more attractive.

Fossil foods deliver **empty calories** where consumers eat lots of calorie-laden food but ingest only a few nutrients per bite. As a consequence, over two billion people globally suffer from micronutrient deficiencies.[92]

The four most prevalent deficiency diseases are malnutrition, nutritional anemia (iron and B12 deficiency), (vitamin A deficiency) exophthalmia and endemic goiter (iodine deficiency). At least half of all children under 5 years of age suffer from vitamin and mineral deficiencies.[93]

The chronic, often lifelong maladies caused by micronutrient deficiencies that occur at different life stages are shown in the diagram.

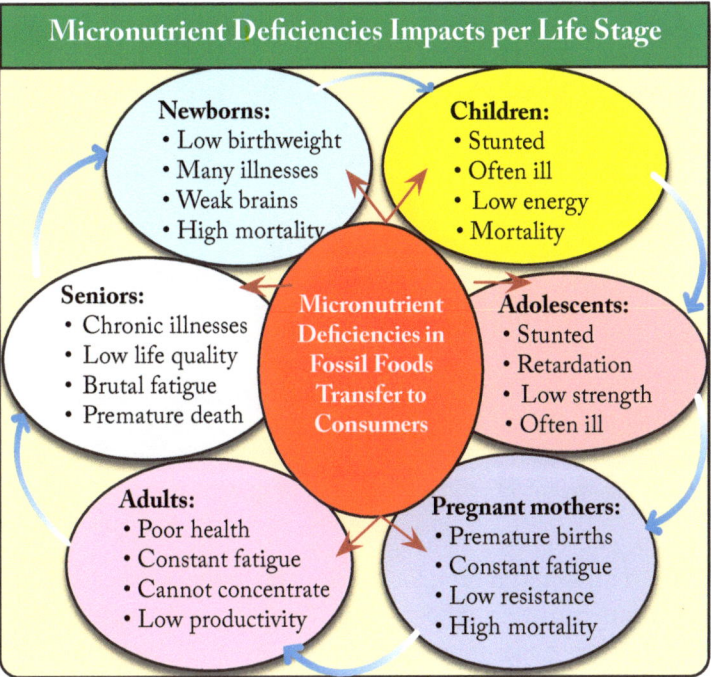

Wealthy countries are not spared as 85% of Americans are nutrient deficient in 2021.[94] These too often result in preterm births which can cause birth defects, developmental disabilities, mental retardation, reduced immunity, blindness, poor learning and premature death. One in six, or about 15%, of US children ages 3 through 17 have a one or more developmental disabilities.[95]

The 380,000 premature births in the US, 1 in 10 births, is higher than any other wealthy country and resembles sub-Saharan Africa. Preterm birth is the most frequent cause of infant death and disability and costs the US healthcare system $26 billion a year.[96]

Half of US adults over 50 suffer from a chronic health condition and over 1 in 4 suffer from two or more chronic conditions. Many chronic health conditions are directly related to the fossil foods diet, which costs US citizens over $3.3 trillion dollars a year.[97]

Life expectancy in the US declined in several consecutive years since 2015.[98] The Western diet contributes to millions of early and middle age disabilities and premature deaths.

IMA focus on yield has led to breeding new varieties that improve yield, pest resistance and claims, (often false), for climate adaptability. Hybrid and GMO crop development has allowed plants to grow bigger more rapidly, but their ability to manufacture or uptake nutrients, especially micronutrients, has not kept pace with their rapid growth.

Junk food addiction

Roughly half of all dietary calories the average consumer eats in 2021 comes from industrial formulations, also known as fast foods. Industrial foods like breakfast cereals, bread, cakes, pies, pastries, cheese and snacks are cheap, convenient and formulated to taste good.

Food marketers, like farmers, are rewarded based on sales volume. Consequently, many have ignored nutrition and health. Food companies have large consumer behavior groups that look for ways to entice consumers into eating more and more their food. Some view food addictions as good because it assures repeat purchases and continuous sales.

Consumers have become addicted to eating processed foods like pizza, pasta, potato chips, cookies, ice cream and cheeseburgers. The ingredients are derived from naturally occurring foods that are stripped of components that slow absorption, such as fiber, water and protein.

Processed foods are engineered for addiction like cigarettes and cocaine. The ingredients are refined and processed into products that are rapidly absorbed into the bloodstream.

Quick absorption enhances the food's ability to light up regions of the brain that regulate reward, emotion and motivation. The most addictive processed foods contain large amounts of fat and refined carbohydrates.

Salt, sugar, oils, fats, thickeners, artificial flavors and other additives in highly processed foods strengthen their attraction by enhancing properties like texture and mouth feel.

Some consumers experience withdrawal when they cut back on highly processed foods. Their symptoms are comparable to drug abuse withdrawal, such as irritability, fatigue, sadness and cravings. Brain imaging studies have found that junk food consumers can develop a tolerance to their sensory pleasure over time. This leads them to require larger and larger amounts to get the same enjoyment.[99]

Ultra-processed fossil foods are largely responsible for 1.9 billion overweight people today. Nearly half a billion people worldwide have diabetes, which causes over 1.6 million deaths each year. The prevalence of obesity and diabetes have been steadily increasing over the past few decades, especially among children.

Chapter 13, Freedom Foods describes how we can flip the deadliest food in human history to the healthiest. Healthy food assures with zero hidden hunger and zero micronutrient deficiencies. Freedom foods deliver superior nutrition and health without extensive processing.

Before proposing biosolutions for healthier food, we will address a significant challenge to human and biodiversity loss – habitat. The **Emerald Forest Initiative** promises to end forest and biodiversity habitat loss in a decade with an aligned set of actions.

Save Biodiversity

6. Emerald Forest Initiative

God has cared for these trees, saved them from drought, disease, avalanches and a thousand tempests and floods. But he cannot save them from fools. – *John Muir*

Farmers are neither fools nor foolish. Farmers are driven to remove trees by their need for feed – food for their families and feed for animals. Only four commodities are responsible for 90% of deforestation: soy, timber, palm oil and beef. Around 90% of deforestation is driven by intensive mechanical agriculture.[100]

What if farmers were able to make a better living growing microcrops on non-crop land? Would compelling personal profit motives eliminate the need to clear forests?

The **Emerald Forest Initiative** plans to save and restore our forests with a novel approach. Current strategies have failed forests. Eco-shaming does not work. Many quota limits and sanctuary policies have not worked. Deforestation continues to increase rather than decrease. The *Guardian* found that our planet loses a football field of forest every six seconds, 24 hours a day.[101]

The Emerald Forest Initiative employs a four-layer strategy to halt deforestation.

1. **Demonstrate and train** people in abundance methods that require neither natural resource extraction nor cropland.
2. **Educate, mentor and support farmers** and their families to learn and use eco-friendly abundance methods.
3. **Sponsor the Emerald Forest Initiative**, an international competition calling for innovators to prove how microcrops can provide farmers with more income and less risk without abusing forests or cropland.
4. **Invent a green alternative to fuelwood** for heating and cooking. We will teach people how to use microfarms to make cooking and heating oil that burns cleanly.

Every person who clears a forest to grow crops knows deforestation is neither sustainable nor a positive legacy for future generations.

Profit motive

Farmers remove trees from forests in search of profit. Farmers believe they can make more money farming cleared forest land than other job alternatives, even if fertility lasts only a few years. Ask a farmer. You will hear how this negative opportunity from a social point of view appears to be the best personal option.

Clearing and farming deforested land forces farmers to endure substantial economic risk. The process imposes heavy physical labour and health risks from heavy equipment and agricultural chemicals and poisons.

The **Emerald Forest Initiative** focuses on saving and restoring our forests with a natural strategy.

Farmer profit

Our forests and biodiversity are in deep destress and deserve reprieve. The Emerald Forest Initiative offers several differentiations from current proposals to save forests. The Emerald Forest Initiative can:

1. Stop tropical and nearly all forest loss to agriculture within a decade.
2. Restore both forest ecosystems and their magnificent biodiversity.
3. Transition forestland farmers to a more profitable means of making a living that requires substantially less physical labour as well as physical and health risk.
4. Create emerald bioproduct substitutes for forest land crops that are driving deforestation.
5. Eliminate the dark poisonous clouds of dust, pollens, fertilizer, agrichemicals and agri-poisons that pollute our shared air, water and ecosystems. These invasive pollutants decimate the brains and lungs of fetuses, newborns and children.[102]

Emerald bioproducts such as soy, oil palm and forest products substitutes are grown on non-cropland with no freshwater, fossil fuels, inorganic chemicals, pesticides or agri-poisons.

Emerald bioproducts

Emerald bioproducts go beyond zero extraction, zero waste and zero pollution.

Emerald bioproduct cultivation actively captures, recycles and reuses waste stream carbon and other nutrients. Bioproducts save forest and cropland because they are cultivated in tanks on non-arable land. Emerald bioproducts are highly productive while they simultaneously clean industrial pollutants from the environment.

Emerald biosystems are also called peace microfarms because they save the natural resources over which countries go to war. They are powered by sunshine, so they do not need fossil fuels. Biosystems can produce biofertilizer that repairs degraded forest and cropland. Advanced microfarms have shown the ability to bring dead soil back to life.

Peace microfarms cultivate bioproducts without fresh water by using waste, brine, brackish and even ocean water for cultivation. Half the water stored on the planet is brine, which is not as salty as ocean water but too salty to be potable.

Emerald biosystems can produce a valuable biproduct – clean water. Communities growing bioproducts in peace microfarms can produce 20% surplus fresh water.

Peace microfarms, above, avoid mined chemical fertilizer by biocycling nutrients from air, water and botanical waste streams such as animal manure. Microfarms use natural methods for pest control, avoiding pesticides and herbicides.

Save Biodiversity

Deforestation cascade

A deforestation cascade acts like an avalanche. The cascade spreads chaos and destruction to the atmosphere, water column and communities across a region that once held beautiful verdant forest. A solution to deforestation needs to address each cascade challenge.

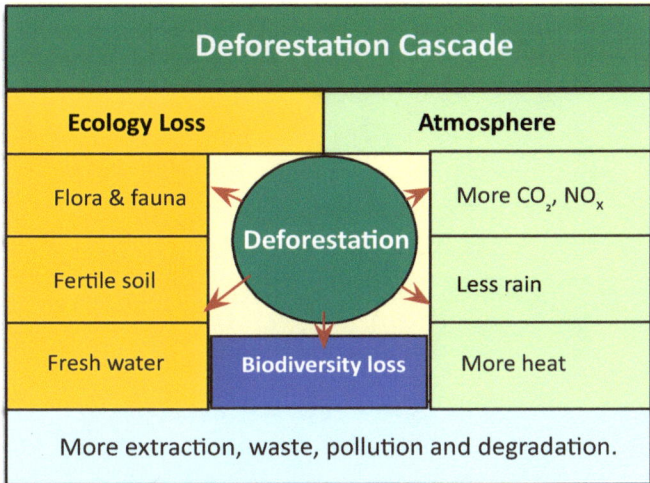

Tropical rainforests are the most diverse ecosystems on earth, harboring more than 80% of the world's known biodiversity. Deforestation causes the loss of 137 plant, animal and insect species every single day, about 50,000 species a year.[103]

Clearing forests eliminates all living things in order to prevent competition with the planned crops. Taking out a forest removes habitat for biodiversity. Clear cutters often spray pesticides, herbicides and fungicides to eliminate pests. These poisons take out just about anything left standing as well as valuable soil microbial communities.

Replacement crops such as soybeans or maize are GMO crops that require more water, fertilizer and pesticides than natural crops. They are genetically optimized to invest their limited energy into maximum seed production. These crops lack the natural defenses needed to compete effectively with weeds, weevils, molds, mildew and insect pests without the help of extra agri-chemicals and poisons.

Deforestation takes a huge bite out of forest lands daily. An area the size of Switzerland (38,300 square km or 14,800 square miles) is lost every year.[104] Cutting and burning of forests releases more than 2.1 billion tons of CO_2 to the atmosphere annually, over 20% of total GHG emissions.[105]

Forests contribute nearly 30% of atmospheric oxygen while 70% comes from algae, largely phytoplankton in the oceans.[106] Destroying forests deprives the region of fresh oxygen while increasing dust, smog and other contaminants.

In some hilly countries, such as Haiti, over 75% of cropland that replaced forests has been abandoned due to erosion.[107] Slash and burn farmers that clear-cut a hillside may find the land holds fertility for only a few seasons. Erosion forces farmers to move on and clear more forest, which continues the cycle of soil loss.

Deforestation disrupts the water cycle, resulting in changes in precipitation and erosion into waterways. Trees extract groundwater through their roots and release it to the atmosphere. When a forest is removed, the trees no longer transpire water. The atmosphere becomes drier and hotter. Deforestation reduces water stored in the soil and severely depletes groundwater. The result decreases moisture in the atmosphere and makes ecosystems hotter and drier.

Chinese scientists found that deforested areas experienced significantly less precipitation.[108] From 1950s to 1980s, precipitation decreased by 33%, making the area in Northern China substantially hotter and drier.

Nearly 75% of the Earth's freshwater comes from forested watersheds. Forest loss degrades water quality and availability. A 2020 UN report on the world's forests found that over half the global population relies on forested watersheds for their drinking water.[109]

Soils and water are under siege. Forests need a minimum of 60% cover to hold the soil and prevent landslides.[110] Replacement crops such as soybeans and palm oil have shallow, weak root systems. They cannot hold the soil.

Row crops such as soy and maize accelerate erosion and overspill, irrigation runoff.

The ecological cascade continues after clearing a tropical forest. Cultivating the soil causes severe erosion to wind and water and adds more CO_2, nitric and sulfur oxide emissions.

Growing crops or grazing cattle adds smog-causing nitric oxides and GHG-heavy methane to the atmosphere.

Beef and oil palm, above, create massive waste streams that foul the atmosphere, water column and all neighboring ecosystems. The manure from animals, often cherished as an organic fertilizer, actually volatizes most of its nitrogen if it is not cultivated into the soil.

Volatization allows the nitrogen to escape and creates noxious nitric oxides in the air, creating smog. In many areas, the waste from meat and dairy animals and oil palm are burned or buried, creating additional dust and soot pollution.

Farmers growing crops on deforested land know they are fighting a battle of diminishing returns. Every year the cost for fossil fuels, fertilizer and pesticides increases while the yields go down due to erosion and soil exhaustion.

Farmers face significant risks farming deforested land. Farming, fishing and timber harvesting are among the top vocations for accidents, disability and death. Heavy equipment creates accident danger. Farmers are exposed to diesel fumes, dust, pollens and insects like mosquitoes. Farming requires day-after-day of hard physical labor. Farmers must have the stamina to complete long workdays.

Farming commodities like soy, maize and oil palm creates economic risk as commodity prices may be below a farmer's production costs. IMA farmers are forced to pay for all the inputs up front. The inputs are used to grow a single crop a year. If the crop fails, the farmers may face bankruptcy.

IMA farmers would like to know a better way to farm and to make a living.

Emerald Forest opportunity

Our World of Data has done an excellent job of describing the primary drivers of deforestation. Soy for cattle feed, oil seeds as food ingredients, field grains and forest products top the list.[111]

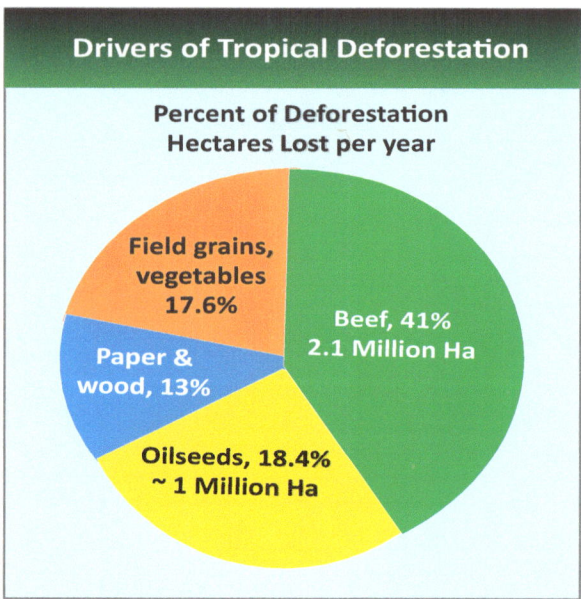

Raising soy or grazing land for beef animals removes 2.1 million hectares of forestland a year. Beef animals are raised on deforested land in Africa, Asia, Latin America and Brazil.

Oil seeds, predominately oil palm and soybeans, consume nearly one million additional forestland hectares a year. Oil palm finds uses in cookies, crackers, cakes, candies, soaps, cosmetics, candles, biofuels and lubricating greases and industrial coatings.

Forest products, including timber, paper and packaging materials are farmed on 680,000 newly cleared hectares each year.

Save Biodiversity

In many areas, people harvest forest wood and dung for cooking and heating fires. Wood and dung fires create massive clouds of black smoke and particulates that cause millions premature deaths a year. The remainder of deforested land is used for field grains such as rice, maize and wheat and vegetables.

Emerald Forest ePrize

The **Emerald Forest ePrize** plans to end deforestation in a decade.[112] The Emerald Forest ePrize will convey one simple message:

Deforestation ends the day forest-land farmers can produce superior microcrop substitutes at lower cost and less risk than soy, oil palm or forest products.

Replacements include better animal feed than soy, better oil than oil palm and superior forest products such as building materials, paper and packaging. These superior bioproducts have already been made from microcrops, but not a cost low enough to compete with heavily subsidized commodities.

The Emerald Forest ePrize will award $100,000 to winners.[113] The Ecolanda website will provide details, including production metrics for palm oil, soy, beef and forest products. Relevant eco-footprint and economic metrics that enable clear comparisons will be posted.

Four Ecolanda Emerald Forest ePrize categories are designed to save forests. Entrants may use narratives, graphics, pictures or videos.

1. **Create a children's booklet** or cartoon that shows how **Anna, the Tiny Mighty Algae**, saves our forests by producing substitutes for soy, oil palm or beef.

Two examples available free here, show how Anna has already saved the world twice.[114] She gave the Earth sufficient oxygen to support life by eating carbon dioxide and exhaling oxygen.

She became the first rung on the food chain. Today algae are eaten by 1,000 times more consumers than any other food because algae deliver superior nutrition. Entrants may use this Anna character, for this competition, or create their own.

2. Share a description of your **microfarm design**. How does your biosystem compare on its eco-footprint with soy or oil palm? (Ten ecofootprint and five economic criteria will be provided.) Standard models will be posted that allow growers to plug in their metrics.

3. Share a description of your **plant-based meat**. How does your process compare on eco-footprint, economics, taste, texture and other dimensions with animal meat?

4. **Create inspiring architectural designs** for future sustainable food cultivation systems. Designs and descriptions may be public art, urban gardens, vertical farms or other ideas.

Emerald Forest Prizes

Grand prizes will be awarded in each category. The top ten entries in each category will be published in order to convey the urgency and innovations designed to end deforestation.

The Emerald Forest ePrize will marshal public sentiment to stop deforestation. We are seeking co-sponsors for the Emerald Forest Initiative. We will work together to communicate how the actions will stop forest and biodiversity loss. Ideal co-sponsors share similar values for halting loss of habitat and biodiversity.

Robert Henrikson and Mark Edwards orchestrated a successful international algae competition. Winners and great graphics are available at AlgaeCompetition.com.

Graphic artists and architects invested over 20,000 hours creating images. This project resulted in a book, *Imagine our Algae Future*, with beautiful visionary public architecture, descriptions of biosystems and fabulous algae-based foods.

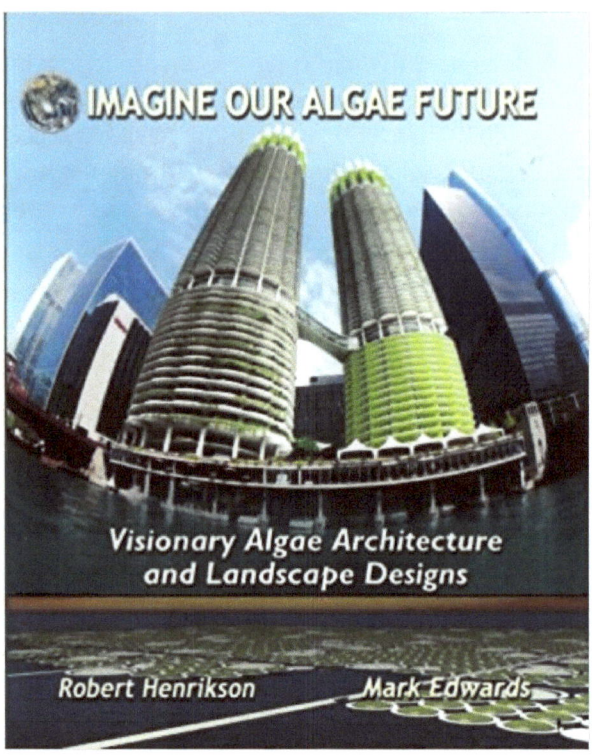

Other Emerald initiatives

Emerald Forest Initiative goes beyond the ePrizes. The competition serves as a form of marketing research to find out how people are attacking deforestation and the progress they have made. We will share this valuable market intelligence globally.

Ecolanda will provide in-person and distance learning for farmers globally to learn abundance methods and how to grow microcrops that save forests and biodiversity. Ecolanda will partner with schools, universities and institutes globally to share eco-friendly abundance methods.

Scientists will assist with training and tools that support ecological restoration and rebuilding forests. A major effort will focus on how to restore biodiversity without harm to sister creatures.

Freedom foods

Nature has been producing microcrop food from microorganisms like algae, fungi, yeast and bacteria consistently for 3.5 billion years. Humans have farmed less than 11,000 years. Only in the last few hundred years have humans destroyed significant forests to produce food.

Farmers can mimic nature and cultivate microcrops using no fertile crop or forest land. Terrestrial field crops grown on deforested land are heavy consumers of extracted natural resources. Microcrops may be called **freedom foods** because they can be grown free of:

- **Cropland,** as they can be grown on rocks, sand or parking lots. Grown in vertical farms, they can grow food more than 50 times more productively per hectare. Therefore, no forests need be cleared.
- **Fossil fuels**, because cultivation is powered by solar energy and other solar derivatives including wind, waves and currents.
- **Freshwater**, since they can be grown in waste, brine, brackish and ocean water.
- **Inorganic fertilizer** as they can receive their nutrients from algae biofertilizer recovered from air, water and biological waste.
- **Pesticides or agri-poisons**, as growers can use natural biopesticides to avoid pests.

Freedom foods use zero extracted resources since they biocycle nutrients. They produce zero waste or pollution as all the nutrients are biocycled repeatedly. They only emission from freedom food cultivation is pure oxygen, lots of oxygen. Every ton of freedom food captures and repurposes nearly two tons of CO_2.

Higher production of terrestrial crops using modern agriculture increases growers' costs and extraction, waste and pollution. Higher production of freedom foods cleans more air, water and biosolid contaminants from industrial production.

A transformation from intensive mechanical agriculture to freedom foods would end deforestation in a decade and begin the process of reversing climate change. Since industrial agriculture uses 92% of the freshwater available, a move to freedom foods would also resolve water abundance problems.

Will farmers adopt abundance methods?

The Emerald Forest Initiative will create proof for farmers that abundance methods are superior to their IMA production. Many farmers will adopt abundance methods to reduce their economic and health risk, as shown in the table.

Farmer Adoption Decision

Agricultural method Food type	Intensive Mechanical Fossil	Abundance Freedom
Economic risk		
High input costs	Yes	No
Crop loss risk	High	Low
Crops per year	One	100+
One commodity	Yes	Multiple
GMO	Yes	No
Profitability	Low	High
Physical risk		
Heavy physical labor	Yes	No
Heavy equipment	Yes	No
Physical stamina	Yes	No
Dust exposure	Yes	No
Agri-chem exposure	Yes	No
Pesticide exposure	Yes	No
Community risk	Yes	No
Children health risk	Yes	No

Farmers will have to be convinced that they can improve their profits by growing multiple bioproducts with abundance methods compared with conventional agriculture. Soy farmers grow and harvest only one crop a year. Abundance growers harvest two crops a week with animal feed that delivers healthier nutrition.

Abundance methods do not need heavy equipment, which reduces physical risk.

Crop loss risk

Field crops are vulnerable to a long list of vectors, some of which are displayed in the hardiness table. Crop loss creates a catastrophe for traditional farmers since they lose all their investment in crop inputs for the full growing season. Field crops like soy and maize are extremely vulnerable to temperature spikes.

A two degree C out of normal range can create a crop loss of 20 to 100%. A temperature spike might slow microfarm production for a few days. When the weather improves, production begins again with harvests twice weekly.

Crop Loss - Hardiness

Agricultural method Food type Hardiness to:	Intensive Mechanical Fossil	Abundance Freedom
Strong winds	No	Yes
Heat spike	No	Yes
Drought	No	Yes
Cold spike	No	Yes
High humidity	No	Yes
Storm	No	Yes
Root attack	No	Yes
Seed/tassel attack	No	Yes
Weed invasion	No	Yes
Insect attack	No	Yes
Fungus and mildew	No	Yes
Need pollinators	Yes	No

Abundance cultivation does require vigilance. Most observations occur with automated sensors and monitors. Data sharing with a central site staffed with experts allow novice growers to make appropriate corrections. Growers make regular adjustments to keep their culture healthy and productive.

Microcrops cultures sometimes crash for a variety of reasons. When this happens, the grower drains and cleans the biosystem and can be producing normally again in a few weeks.

Aquaculture dilemma

Due to vast overfishing, nearly 90% of global fish stocks are either fully fished out or overfished. Humans have managed to wipe out 92% of the ocean's largest fish.[115] These animals: sharks, Bluefin tuna, swordfish, marlin, and king mackerel, are at the top of the marine food chain.

Fishermen now overfish further down the food chain, depleting the oceans of prey fish. Scientific studies predict that aquaculture will outgrow the supply of fishmeal by 2023.[116] Farmed fish production outstripped beef production for the first time in 2015.

The FAO report, *The State of World Fisheries and Aquaculture*, demonstrates that global fish consumption per capita has reached record-high levels due to aquaculture and consumer demand.[117] An average person now eats roughly 44 pounds of fish per year, which doubles the 1960s sea food consumption.

While fish stocks are plummeting, consumers accurately view fish as healthier protein than other meats, which creates an exploding demand for fish protein. Increased demand for fish follows the compelling health benefits research on fish oil, omega-3 fatty acids.

The **Emerald Marine Life Initiative** will focus on saving the small fish in the sea by promoting eco-friendly, algae-based Omega-3 oil to replace omegas sourced from fish.

Today, more fish are grown in aquaculture, 105 million tons) than are caught at sea, 95 million tons).[118] About 70% of forage fish like anchovies, herring and sardines are caught and processed into fishmeal and fish oil for farm-raised species.[119] About 25 million tonnes of wild-caught fish are fed to aquaculture fish annually.[120]

Ecolanda will cultivate algae and other microcrops with superior protein and other nutrients for aquaculture. This will create an industry that will eventually make harvesting forage fish obsolete. Ecolanda plans to cultivate 100 million fish a year with land-based emerald aquaculture.

Microfarm fish feed

An encouraging development for fish feed was announced in 2021. The Canadian company Smallfood discovered a new strain of microalgae that grows to mature biomass with 85% protein in only seven days.

The algae-based fishmeal substitute can be grown on non-cropland and delivers a more balanced protein than animal meat. It emits 30-times less GHG than beef and reduces water consumption by 160-times compared to farmed fish. Current production methods require expensive fermentation equipment. When photosynthetic biosystems can produce protein this quickly at lower total cost, microfarmers will become profitable suppliers of aquafeed.

High GHG emitting materials

Another Ecolanda climate change strategy will address high GHG emitting materials such as cement, asphalt, bioplastics and construction materials. Ecolanda plans to replace these eco-dirty products with clean emerald substitutes that pull pollutants from ecosystems and are biodegradable.

Ecolanda climate initiatives will demonstrate to the world that microcrops offer superior climate solutions and abundance methods are safer and more profitable for growers. Abundance methods are healthier for growers, their communities and the environment.

Forest product substitutes

Industrialized countries consume 12 times more wood and wood products per person than the non-industrialized countries. Almost half of the world's timber and 70% of the paper is consumed by Europe, the United States and Japan.

Ecolanda plans to cultivate superior substitutes for wood and paper products. The paper industry is fourth largest in producing GHG and among the leading industries degrading water. Paper products contribute substantially to deforestation and the related economic, social and environmental costs.[121]

Ecolanda will use the considerable fibres in algae to cultivate paper products that provide better solutions for reading, writing, wrapping and packaging. Packaging production consumes huge amounts of energy, water and other natural resources. Packaging creates billions of tons of waste materials that pollutes our air, water, soil and oceans. By 2050, waste plastic will outweigh fish in the ocean.[122]

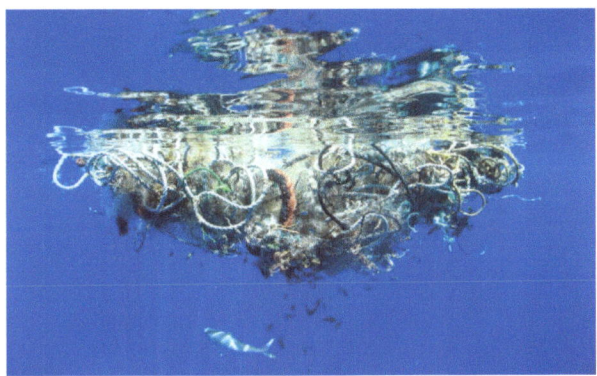

Microcrop substitutes

Algae-based substitutes – cardboard, biofilms, paper and bioplastics – may serve as superior biodegradable replacements that are made with zero emissions or pollution.

Buildings consume massive amounts of timber which increases GHG emissions and diminishes forest stocks. Ecolanda will invent a suite of biodegradable building materials with superior strength, usability and construction attributes compared with wood.

Construction-grade algae biofoam offers three times the tensile strength of wood and twice that of concrete block. Biofoam allows buildings to be constructed in half the time as wood framed buildings. The material makes buildings bug, odor and earthquake proof as well as fire resistant. Biofoam delivers superb insulation and cuts lifetime utility costs by 80%.

Algae-based construction material substitutes currently are not cost competitive. Ecolanda will improve productivity, which will change the economics. When the economics flip, biodegradable packaging and construction materials made with abundance methods will dominate construction markets.

Fuelwood

People in many countries cut and burn firewood several times faster than tree growth rates. Families must constantly travel further into the forest to find wood. Some families spend most their day searching for firewood.

Harvesting firewood indiscriminately expands deforestation and reduces timber resources. Timber reduction creates loss of habitat and leads to extinction of the species that once lived in the habitat.

Half the people on earth tonight will cook their supper over an open fire. Most cook over an open fire of wood, animal dung or coal in a small, enclosed kitchen.

Household air pollution causes stroke, heart disease, asthma, chronic obstructive pulmonary disease and lung cancer.[123] Over 7 million people die prematurely every year from black smoke pollution from smoky cooking and heating stoves.

About half the smoke deaths are caused by pneumonia among children under five. Millions more children and often their grandparents are disabled by black soot particulates inhaled from household air pollution.

Ecolanda will cultivate clean burning cooking and heating stove oil. Algae oil provides higher heat per kilogram than wood. Algae oil is easy to use and burns cleanly, with no black smoke particulates. Algae oil emits no black smoke because the algae did not go through fossilization.

Most petroleum, coal, shale, and natural gas products are derived from fossilized algae from ancient lakes.

Algae oil does give off a light odor similar to French fries.

SCAD Ecolanda will train farmers and families how to use abundance methods in microfarms to grow cooking and heating stove oil, biofertiliser and animal feed. Local training will provide independence and health to many in Ecolanda's neighboring communities.

The Emerald Forest Initiative will discover natural microcrop substitutes for forest products from global microfarmers competing for ePrizes. Forest product substitutes will allow regions to reforest and their colourful biodiversity to re-establish and flourish.

The next section examines a novel opportunity to train farmers in abundance methods.

Elephants are the poster animal for ZooPoo

7. Emerald ZooPoo Initiative

Algae transform poo at the zoo to clean, green bioenergy products for animals and plants. Biocycling nutrients for reuse benefits the animals, plants, zoo and biodiversity.

Nature has used biological regeneration sustainably for eons. Nature recycles and reuses nutrients many times. Nature wastes nothing. One life supplies the nutrients for the next. Nature biocycles and regenerates life. Nature's basic principle eliminates waste:

Your waste is my food.

Waste carbon and nutrients from zoo waste streams become the food energy that algae pair with solar energy to grow valuable bioproducts.

The key to successfully following nature's path requires thoughtfully integrating carbon and other nutrient pollution sources and syncing them with algae biofactories designed to capture nutrients and to grow bioproducts.

Rather than positioning nutrient recovery and reuse on a farm, where only a few people can see it, why not create a world class demonstration project at the zoo, where millions can share and see the experience? If we can do ZooPoo at the zoo, we can recover and repurpose energy and nutrients anywhere.

Futurists have positioned algae bioremediation and bioregeneration for NASA and DARPA's 100-Year Starship project. Algae are proposed to provide food, life-support, waste cycling, green chemicals, bioenergy, and building materials. Algae are also recommended to make the medicines necessary for 100 years of life in space. ZooPoo provides a more down-to-earth opportunity for algae biosolutions.

The intriguing positioning of algae cycling and reuse in both a long-voyage spaceship and a zoo creates an important question. NASA calculated that six liters of algae water will produce daily:

- 600 grams of food, (540 grams yields 2500 calories, an average daily food requirement)
- 600 liters of oxygen
- While consuming 720 liters of CO_2.

Is there any other organism on earth that can sustain both a starship and a zoo?

ZooPoo
Recover, recycle and reuse energy, water and nutrients from the zoo waste stream.

Imagine that your zoo becomes the world's first EcoZoo and demonstrates nature's preferred mode of energy storage and harvest – green solar energy in rich algae biomass.

Bioenergy system

Nature's first and simplest bioenergy system, algae, uses only sunshine, wastewater and surplus CO_2 to recycle and reuse ZooPoo to produce clean, sustainable, carbon neutral food, feed, energy, fertilizer and freshwater.

ZooPoo allow a zoo to move to a **net-zero**:
- **Carbon footprint** – zero CO_2 emission
- **Freshwater footprint** – extra freshwater over animal and human use
- **Fossil fuels footprint** – zero fossil fuels
- **Fossil nutrient footprint** – zero extracted nutrients or fertilizer
- **Pesticides or agri-poisons** – none, clean
- **Eco-footprint** – positive eco-footprint

The ZooPoo exhibit shares an important story:

> **Farmers** *may grow healthy foods and bioproducts while protecting all the animals and plants on earth from the overconsumption, waste and pollution associated with industrial agriculture.*

Zoo waste problems

Discussions with zoo directors and managers revealed that zoos often pay more to manage ZooPoo, botanical and animal wastes, than their animal feed. In the name of ecological safety, bureaucrats have levied layers of protections, e.g. inspectors from various bureaus, on zoos to monitor their waste management. California imposes no less than five agencies to monitor ZooPoo waste streams.

Zoo waste management has become a major cost because the current process requires extensive labor and considerable energy. Zoos must carefully gather their waste products and transport them to a holding facility where no discards can leach into the ground – even if it rains. Typically, the poo (animal manure) must be covered to avoid unpleasant odors. After inspection, the manure wastes are loaded into trucks and transported long distances to an approved waste dump. Not all waste dumps are equal.

In addition to all these costs, local or state governments often add fees by weight to "exotic" waste that go to dumps. To date, no one has explained how exotic animals like elephants' poop differs from fido or human waste.

Incredibly, zoos must go to substantial expense to manage ZooPoo. Then the zoo forfeits the entire poo energetic nutrient value to a costly, often distant and expensive waste dump.

ZooPoo value

ZooPoo retains significant value. ZooPoo contains roughly **60% of the energy** originally in the plants eaten by zoo animals. Elephant poo contains about 93% of the original plant energy. Elephants have the biggest appetites of all zoo animals, but they have the lowest energy and nutrient absorption in their large stomachs and intestines.

Elephants are enormously rich poo biofactories. Zoos provide an excellent model for nutrient biocycling because ZooPoo retains **80% of the nutrients** originally in the plants eaten by zoo animals. Zoos currently lose both the energetic and nutritive value in animal and botanical wastes.

Each zoo dumps a gold mine of rich energy and recoverable nutrients into already overfilled waste dumps. ZooPoo bioremediation and bioregeneration can transform a huge zoo cost to a profit centre. Even better, the process can create a superb destination for ecotourism and learning centre for abundance farmers.

Kill waste dumps

Eco-smart states, such as California, Massachusetts, Oregon, and Washington have passed [disposal bans](#) on certain products, including yard materials and botanical waste. Those wastes are removed by vehicles on a fee-for-service basis. Experience shows there is no effective practical way to enforce a recycling goal. However, public officials are able to enforce disposal bans. As a political strategy, pursuing disposal bans has a better chance of success with public acceptance than attempting to pass a comprehensive state or federal solid waste plan.

Many countries in Europe and central America have passed laws to completely prohibit dumping. The motivation for these dumping bans is simple; countries have run out of land acceptable for waste disposal. Spain, Denmark, Canada and other leading green countries are finding surprisingly profitable business in waste-to-energy technologies, including gasification and gas plasma.

Conventional waste-to-energy plants that use mass-burn incineration can convert one ton of MSW, (municipal solid waste) into about 550 kilowatt-hours of electricity. Gasification technology is about twice as efficient with feedstock and uses one ton of MSW to produce about 1,000 kilowatt-hours (kWh) of electricity.

Plasma gasification offers similar efficiencies and can be used to convert carbon-containing materials to synthesis gas that can be used to generate power directly or products that store energy like hydrogen. One liter of hydrogen delivers three times the energy of gasoline.

Gas plasma provides safe waste-to-energy solutions for hazardous, biomedical, hospital and chemical materials. Anaerobic digestion (see discussion below), of wet organic wastes can generate 300 to 1,000 kWh per ton while producing digestate as a valuable byproduct that can be composted as a soil amendment or further processed into solid and liquid fertilizer for growing beneficial crops.

Mass-burn waste-to-energy technologies create two pollutive emissions, incineration flue gasses and the biochar residual in the combustion chamber. The flue gasses can be remediated by algae to produce more biomass. The biochar serves to safely sequester potential pollutants in a slow-release soil amendment to support crops.

ZooPoo goals

The ZooPoo exhibit architecture will demonstrate the value of nutrient biocycling. Many zoo guests recycle their organic material at home and their grocery bags. They are already aware of the recycle-reuse value proposition.

ZooPoo transforms the substantial zoo waste management cost to a profit centre. Target market guests, those who currently pay to burn or bury their wastes, include:

1. **Farmers,** who will be able to realize value from animal and plant wastes and moderate pollution and waste.
2. **Municipal waste facilities,** that will be able to create value from human waste streams.
3. **Community (food waste, garden and trash) waste facilities,** that will create value from recyclable trash and garden botanicals.
4. **Power and cement plants** and manufacturers, that will create value from their surplus CO_2 while avoiding emissions.
5. **Citizens** who desire to learn how to minimize their waste streams and ecological footprint.
6. **Children** who wish to convey conservation and renewal to their peers and parents.
7. **Churches** that desire to convey green and sustainable lifestyles to their communities.
8. **Schools** that want engage students in sustainable systems.
9. **Green and environmental social networks** that want to see ecologically responsible production of food, feed, and freshwater.
10. **Zoo visitors** interested in food and energy security by reclaiming and recycling surplus inputs that are affordable and, unlike fossil resources, will not run out.
11. **Visitors committed to global stewardship** by moderating pollution while producing valuable bioproducts that biocycle or store rather than release carbon.
12. **Educators** in sustainable and affordable food and energy (SAFE) production, that creates a positive ecological footprint.

ZooPoo can be designed to convey mission critical solutions for social, economic and environmental challenges.

Sustainable You at the Zoo

Many animals, plants and entire ecosystems are threatened with extinction due to global climate change, agriculture expansion, food production, freshwater scarcity and the availability of fossil fuels and fossil nutrients, including fertilizers.

ZooPoo addresses each of these challenges and shows visitors how to change their own behaviors to save animals, plants and ecosystems as they enhance our communities. "*Sustainable You at the Zoo*" will provide a take-home checklist to support ecosmart lifestyles.

ZooPoo will show energy and nutrient recovery, recycle and reuse demonstration. Guest will see, learn and experience how to adopt green behaviors and lifestyles.

ZooPoo uses the zoo waste stream, ZooPoo, and recovers and reuses energy and nutrients to produce electricity, freshwater, vitamins, minerals, health foods, animal feed, fertilizers and medicines for zoo animals and plants.

Algae recover the hydrocarbons stored in ZooPoo, which may be used for energy rather than burning fossil fuels. The facility will also include demonstrations of other renewable forms of energy such as solar, wind and possibly other means such as geothermal systems.

Elephants are among the smartest animals at the zoo. They may serve as the ZooPoo icon because each adult elephant contributes more than 300 pounds a day to the ZooPoo pile.

Elephants have the most inefficient digestive systems, at 40% of their intake, even though they have 19 meters (21 yards) of intestine.

ZooPoo will demonstrate how to transform ZooPoo from waste to valuable bioproducts that benefit plants, animals and people.

Cycling history

Mother Nature has been cycling nutrients successfully with algae for 3.5 billion years. ZooPoo builds on nature's first and most efficient food and energy production system.

Many people are unaware of the potential for algae and other microcrops to provide carbon neutral food, feed and fuel. The few algae producers have worked largely under the public radar, distant from population centres.

ZooPoo raises algae farming to a new level by enabling algae production and its many coproducts from the zoo's waste stream. After constructing the demonstration, production, and education centres, ZooPoo will be self-sustaining environmentally and economically.

Algae production creates biomass with surplus inputs that are cheap and will not run out – sunshine, CO_2 and wastewater nutrients. Algae production systems get most their needed energy free from sunshine but require some additional energy for mixing and extraction.

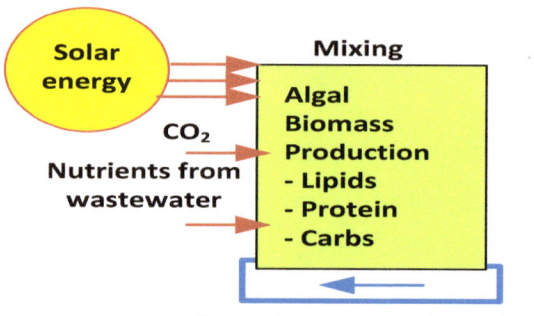

The energy stored on earth comes from the sun. Nature transformed the algae biomass from ancient oceans into fossil fuels, but the process took tremendous pressure and heat over 400 million years. Fossil fuels offer a convenient form of concentrated energy, but pollute the atmosphere with heat-trapping gases, as well as heavy metals and black soot particulates. Algae produces biofuel in weeks rather than eons.

Algae biofuels displace the use of fossil fuels gallon-for-gallon. Unlike fossil fuels, during production, algae cultivation releases only pure oxygen to the atmosphere. Algae oil creates clean, renewable biofuels that burn with no black soot particulates because the algae oil has not fossilized. The ZooPoo facility will have a small motor running on algae oil. Guests can attest that the simple vegetable oil burns cleanly but has a hint of an odor similar to French fries.

After recovery of algae oil for energy, other coproducts may be extracted from the remaining biomass. or can be processed into renewable natural gas or transportation fuels. Solid wastes are remediated through anaerobic digestion that provides both gasses and wastewater that provide more algae feedstock.

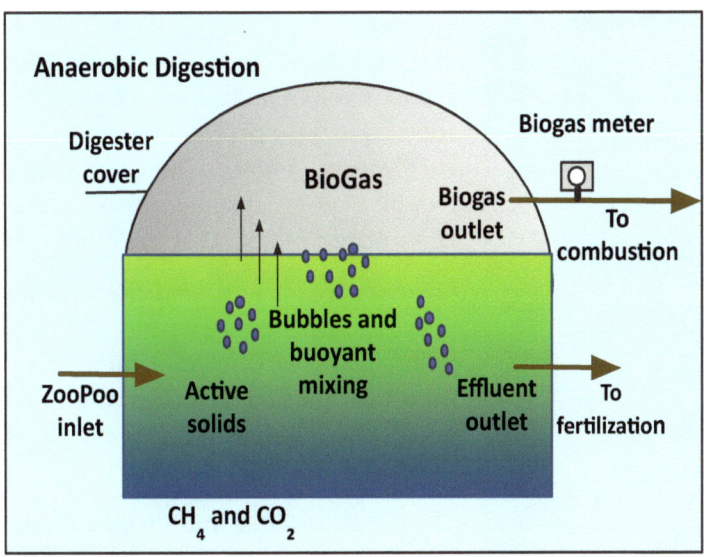

Anaerobic digesters use a series of biological tools and apply micro-organisms (anaerobes) to break down biodegradable material in the absence of oxygen. An end product, biogas, may be combusted to generate electricity. Flue gas from combustion can feed more algae.

The water used to culture the algae may come from any source, but high-nutrient wastewater offers many advantages, especially free nutrients. Water may be recycled, and the residual nutrients reused up to 10 times. Lipids may be pressed out of the biomass for use healthy oils for animal feed or converted to liquid transportation fuel.

Algae protein provides food energy for animals, fowl and fish. The remaining carbohydrates can be refined into energy, biodegradable bioplastics, paper, fabrics, green chemicals and many other bioproducts. Algae can also provide vitamins, medicines and vaccines for the zoo animals.

The ZooPoo exhibit purpose is simple: show rather than tell visitors how to recover, recycle and reuse nutrients from zoo animals and botanical wastes. The message also includes how to practice abundance methods that use no or minimal fossil resources.

ZooPoo Process

ZooPoo includes the liquid and solid the wastes from animals, plants and zoo trash. The exhibits will use about 60% virtual – videos, and 40% interactive. Waste-to-energy videos will provide guests with sights of amazing processes using big screens that avoid the required waste footprint, odors and noise. The interactive exhibits will entertain and engage target guests of all ages.

ZooPoo employs a clean and adaptable carbon-neutral production process that consumes large amounts of CO_2 and transforms the carbon into high-value products while releasing pure oxygen into the atmosphere.

ZooPoo uses abundant, (renewable) or surplus inputs while producing valuable bioproducts and pollution solutions. ZooPoo biosolids will be filtered, dried in the sun and then burned in a closed kiln in a process called pyrolysis. Burning the organic matter in the closed system releases no CO_2 to the atmosphere and creates three valuable components: H_2, CO gases and biochar.

The H_2 is piped to a generator to create more electricity for the ZooPoo exhibit. The carbon monoxide gas is piped to an algae biosystem where it provides the carbon for additional algae biomass. The biochar becomes a slow release biofertilizer and soil conditioner for zoo trees and plants.

The core carbon negative technologies, provide a showcase for abundance. Carbon negative production means that carbon is removed from the air. Abundance production methods ensure that no or minimal fossil resources are consumed in production by using several biotechnologies.

ZooPoo — Recover, recycle and reuse the zoo waste stream

Inputs: Solar energy, CO_2, Wastewater, ZooPoo

ZooPoo includes liquid and solid waste from:
• Animals
• Plants
• Zoo trash

Algaculture – cultivated algae creates:
• Lipids (oil)
• Protein
• Carbohydrates

while cleaning the water.

Mixing — Recycle culture and nutrients

Products and pollution solutions (Harvest):
- Feed for animals, birds and fish
- Food ingredients
- Carbon sequestration
- Clean water, fresh air and pure O_2
- Organic fertilizer
- Vitamins and micronutrients
- Fine medicines and vaccines

No or minimal use of:
• Cropland
• Freshwater
• Fossil fuels
• Herbicides

Nutrient recovery – algae bioaccumulate and store nutrients from the poo tea created from animal and plant waste streams.

Energy recovery – burning animal and plant solid wastes in a closed system, gasification or pyrolysis, creates H_2 for energy production and more CO and CO_2 to feed algae.

Clean water – algae clean wastewater and make it suitable for irrigation, animal or human use. Wastewater treatment is the oldest algae application in the US and dates back 90 years.

Energy production – algae are harvested, and the oil is pressed out to create clean, green diesel that burns with no black smoke particulates. Additional energy is created by the H_2 produced from gasification of solid wastes.

Biofertilizer – selected residual from algae production may be used as a fertilizer for the many plants at the zoo. Biochar, a byproduct from pyrolysis waste management, provides rich biofertilizer and soil amendments for zoo plants. Some local algae will be grown as liquid biofertilizer for drip systems in greenhouses and in hydroponic vegetable production.

Biofeeds – high protein algae cultivars will be grown for animal feed. Most animals will receive a mix of algae and food grains, or in the case of carnivores, meat. The aquaponics exhibit will display fish eating repurposed nutrients in their diet.

Farmaceuticals and medicines – algae compounds are harvested to make animal health foods, vitamins and minerals, including Omega-3 fatty acids, which improve the health of zoo animals.

The ZooPoo algae production system illustrates the steps in carbon neutral production where fuel, electricity, food, feed, fertilizer, and other products are made using no or minimal fossil fuels or fossil carbon products such as fertilizers or agricultural chemicals.

Algae may be harvested and processed to create a wide range of bioproducts. Each algae bioproduct is biodegradable, including building materials. The ZooPoo exhibit will demonstrate not only the materials but also explain how biodegradable construction and packaging materials benefit our society and environment.

Excrement

ZooPoo may not be pretty in a classic sense, because waste streams are dirty. The ZooPoo mantra will show in various ways how recycled nutrients can create bioproducts that are clean for people, animals, plants and our environment. Critically, the exhibit will smell good because released orders indicate a waste of valuable nutrients. The exhibit demonstrates abundance methods that show how to eliminate waste.

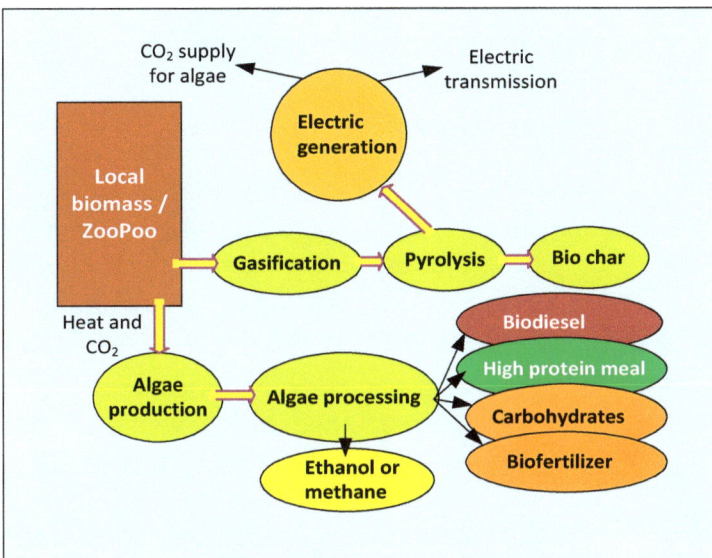

ZooPoo will turn poo upside down and celebrate poo as a treasure that transforms brown into gold; clean, green food and energy. Exhibits will demonstrate steps in the process to assure cleanliness, healthiness and nutrient value.

There will be many examples of how poo is used globally, such as China's farmers that have used human poo on their fields successfully for thousands of years. Organic farmers use animal waste regularly on their fields to produce healthy crops.

Students may not know that beer and wine come from the excrement of yeast cells, or that the Plains Indians, as well as early settlers used buffalo poo regularly for their cooking fires. Poo retains high-value energy and nutrients, why waste it?

ZooPoo Exhibit

ZooPoo will benefit zoo visitors of every age. The learning facility will serve as a gathering point for environmental and socially conscious networks. ZooPoo provides engaging learning opportunities for people interested in carbon negative, Emerald, production of food and energy. People will enjoy ecologically sensitive lifestyles as well as water, food and energy conservation. The facility will support extensive R&D plus visual, interactive exhibits for sustainable and affordable food and energy.

The ZooPoo exhibit will cover a hectare. Exhibits will include Peace Microfarms growing algae for several purposes, a microorganism interactive learning centre, as well as several satellite exhibits.

A consortium of experts will design the exhibits. We will engage environmental design architectural students, as well as professional architects to develop ideas for the learning facility. We will also rely on the Zoo Board of Directors to align the project with the vision and values embedded at each zoo.

There will be interactive displays about the miracles algae have already created. Displays will show how algae were used by ancient cultures for food, feed, fertilizer and medicines. Of course, there will be plenty of nutritious, tasty and free algae-based foods to sample.

Learning centre

The microorganism interactive learning centre will bring microbes to life on big screens. Guests will be able to see live microbe interactions including feeding, mating, birthing offspring and the many special things microbes do. Selectable videos will provide entertaining microorganism education. Some lucky guests will be able to perform algae bioprospecting right on site. If they find an organism not already named, they will be able to choose a name for the new organism. Advanced sensors paired with big data and artificial intelligence will assist in the search for new microbes.

The algae biosystems may use bags, tubes or flat plastic rectangles as illustrated in the pictures available at Imagine Our Algae Future. The microorganism interactive learning centre will collaborate with Micropia in Amsterdam, the Smithsonian, the Exploratorium in San Francisco and imaginative STEAM, (science, technology, engineering, arts and math) learning sources such as the Arizona Science Center.

The signature ZooPoo exhibit will display the benefits of Abundance farming methods and Freedom Foods, based on the power of single-celled organisms. The context will be a spaceship traveling for years to distant planets. The astronauts will have to grow their own sustainable food, feed and fertilizer. The interactive display will allow student astronauts to cultivate bioproducts they want for their journey to distant planets with algae.

From Micropia in Amsterdam

After student astronauts understand how nano cells can produce sustainable food in a spaceship, their transition back to earth will seem quite simple. They will follow Ana's journey to create food justice by recycling nutrients and repurposing them for zoo animals, food crops and people. Students will be able to use 3D printers to print a variety of healthy algae-based foods. They will be able to design favored tastes, colours and textures into their food products.

Guests will learn how animal production over-consumes finite resources, uses them once and then pollutes ecosystems. Animal production consumes tremendous amounts of resources and adds substantially to greenhouse gases.

Students will see cows happily growing in pole barns with peaked roofs. Fresh air comes in the sides of the barn while the cows' substantial methane emitted from both ends of the cow, rise to the peaked roof. When a sensor identifies a sufficient concentration of methane, a pump flows the gas to a combustion chamber, where it is burned for energy that powers the farm's microgrid. The combustion chamber gasses are piped to an algae raceway to provide carbon for the next algae generation.

ZooPoo summary

ZooPoo focuses on two primary objectives.

1. Share the value of the zoo waste stream to demonstrate how energy and nutrients can be recovered, recycled and reused for a wide variety of products and solutions.

2. Motivate and train farmers to adopt abundance methods and biosystem solutions in place of intensive mechanical agriculture.

ZooPoo will demonstrate human future lifestyles in terms of sustainability, conservation, pollution solutions and ecological preservation. ZooPoo will provide practical demonstrations of solutions desperately needed by current societies including how to produce sustainable and affordable carbon neutral food, feed and fuel without using fossil resources.

Several exhibits will demonstrate the substantial value proposition for algae biofertilizer for plants and biofeed for all types of animals. These exhibits will emphasize superior nutrition as well as the substantial ecological advantages.

The next section, the Ecolanda Emerald Bioeconomy provides a demonstration site to provide the critical metrics to support abundance methods.

8. Emerald Bioeconomy

Ecolanda provides a global demonstration site for an emerald circular bioeconomy.

Ecolanda achieves a sustainable circular bioeconomy due to an architecture based on nature, smart biotechnology and 7-generation stewardship.

A circular economy uses renewable bioresources from land and sea. Bioresources include crops, forests, fish, animals and micro-organisms to produce food, materials, consumer products and green energy. Most bioeconomy projects are defined as "green" and hope to reduce a carbon footprint sometime in the future.

Ecolanda will be the first **emerald bioeconomy**. The architecture connects every system to assure the sustainable cultivation of superior goods with minimal extraction and zero waste or pollution. Ecolanda is built to be carbon neutral on day one.

Ecolanda's emerald bioeconomy goes far beyond carbon neutrality by producing superior bioproducts in an eco-responsible manner that:

1. Are **carbon negative**, capturing and reusing carbon and sequestering carbon.
2. **Integrate smart biosystems** that clean and restore air, water and ecosystems while cultivating valuable bioproducts.
3. Apply **7-generation stewardship**.

Each Ecolanda project plans to capture and reuse over 75 million metric tons of carbon annually.

7-generation stewardship.

Ecolanda evolved from 40-years of global agri-energy sustainability research. The principal investigation focused on:

Will the proposed process operate cleanly and smoothly for 7-generations?

Indigenous people have much to teach us. Native people lived for centuries in their sustainable communities. Indigenous people revered nature – air, water, land and wildlife.

Native communities often made decisions based on 7-generation stewardship. Long before the current attention to sustainable systems, Native people asked what impact their decision today would have on the welfare and wellbeing of children born in 7-generations, 140 years.

Ecolanda uses 7-generation stewardship to ensure positive impact on nations and communities. Several Ecolanda projects are planned on four continents. Each takes the name of the host Nation or region and selects clean biotechnologies based on 7-generation stewardship.

Idyllic cities and communities have been proposed through history. A few have been built, such as Gaviotas, a mountain village in Colombia, South America.[124] Most ideal communities were based on impractical theory, not reality.

Foundation for Ecolanda

Ecolanda benefits from a solid foundation built primarily on tested biosolutions that exist today. Dozens of site visits, hundreds of interviews and many techno-evaluations drove the selection of eco-friendly solutions. Many excellent tools are available today but are not sustainable as stand-alone systems.

SCAD, Southern Cross Agri-energy Development

SCAD serves as the Ecolanda systems integrator and links clean technologies into circular biosystems. SCAD's analysis considered and tested the pros and cons for mechanical and biotechnologies that claimed agri-energy sustainability.

Several conclusions emerged:

1. Extractive methods are sustainable only until the first needed substance goes extinct or extraction becomes too expensive. Extractive processes like industrial agriculture will leave our future children without the precious natural resources they need.

Prior generations did not concern themselves with resource scarcity because resources were plentiful.

Today, only 70 years since the start of the Green Revolution in agriculture, many of the valuable natural resources needed for food and fibre production are past their peak extraction and are moving rapidly toward extinction. Soil nutrients, fresh water and phosphorus are prime examples.

2. Industrial mechanical methods are not sustainable. Continuous extraction, waste and pollution add expense and destroy ecosystems. Circular biosolutions avoid extraction and waste and are sustainable.

Farmers have gone bankrupt and left their farms in many communities due to weather, water or the increasing cost of agri-inputs. Extraction continually becomes more expensive as the quality of ore or target substance diminishes and refining becomes more difficult.

Systematic nutrient extraction from cropland has left more cropland exhausted and abandoned than is farmed today.[125] Most good cropland has been farmed for decades. Farmers search for new cropland but find the land less flat, less fertile and substantially more difficult to farm. Cropland expansion drives deforestation which contributes 20% of annual GHG to the atmosphere.[126]

3. Eco-friendly methods may not sustainable when they operate alone.

They need the substantial benefits that accrue from systems integration. Integrated systems link each process with upstream elements for inputs and downstream systems for waste capture and nutrient reuse.

4. Industrial mechanical agri-energy solutions are not long-term viable.

Agri-energy solutions need to be designed to resolve the substantial risks from climate change. Agri-production must stop the systemic degradation of our atmosphere, cropland, waterways and ecosystems.

The top five global meat and dairy firms are now responsible for more annual greenhouse gas, GHG, emissions than Exxon, Shell or BP.[127] Industrial agriculture also is responsible for over 80% of hazardous water pollution.

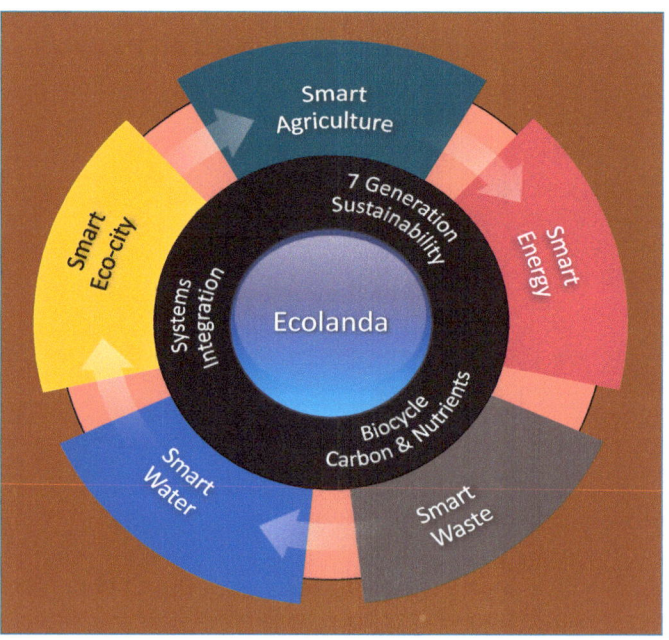

5. Sustainable solutions must not only avoid adding more environmental pollutants but also remediate toxic wastes accumulated from decades of industrial production.

Our children and their children deserve clean air, water, lush ecosystems and oceans filled with rich marine life. Biosolutions to attain these goals are examined in the following chapters.

National Stewardship

National leaders in the host country shape each Ecolanda project. Ecolanda planning commences with the host country selecting a National Stewardship Team. Ecolanda typically represents the largest private investment in a country's history, €15 Billion. Joint stewardship and collaborative planning develop strong National goals, timelines and action steps.

Integrated biosolutions

Ecolanda demonstrates how integrated biological systems, circular economies, create significantly higher value compared with industrial mechanical methods. National value metrics report benefits in health, economics, society and environment.

Residents and neighboring communities will embrace Ecolanda due to improvements in health, affordable and plentiful food and living standards.

Ecolanda serves ecotourists, students, teachers and scientists with fascinating demonstrations of integrated sustainable systems. Ecolanda's architecture creates the highest agri-energy productivity while capturing carbon.

SMART biosystems

SCAD employs **SMART** integrated biosystems. These **S**ustainable **M**icro **A**bundance and **R**egenerative **T**echnologies are each sustainable and non-pollutive for 7-generations.

Micro Abundance methods mimic nature and create superior food and other consumer and industrial products in a circular bioeconomy.

Nature has grown microcrops consistently daily for 3.5 billion years. Trillions of hungry consumers eat microcrops daily because they typically deliver over twice the nutrition per bite as land-based crops. Small creatures such as shell and finfish depend on algae for food. Large animals including giant blue whales get their primary nutrition from algae and algae feeders like krill.

Microcrops such as algae provide 40% of the new biomass on earth daily. These microscopic plants produce 70% of the planet's new oxygen daily.

Microcrops enable growers to cultivate a wide array of valuable bioproducts without consuming precious fossil resources. Abundance growing methods provide significant advantages compared with environmentally destructive industrial mechanical agriculture. Abundance methods allow growers to:

- Biocycle waste stream nutrients from air and water, which reduces input costs and avoids natural resource extraction.
- Use minimal or zero cropland, fossil fuels, fresh water, chemical fertilisers, pesticides or other agri-poisons.
- Create zero waste, GHG emissions and pollution. The only emission is pure oxygen.
- Clean air, water and ecosystems while producing valuable food, consumer and industrial bioproducts.

Regenerative technologies

An end to the enormous extraction, waste and pollution from mechanical agri-energy production provides substantial value. However, nature works slowly and would require centuries to remediate existing environmental contamination.

Ecolanda goes much further than the "Do no harm" command with biosolutions that accelerate nature's ecosystem recovery.

Regenerative technologies play a critical role in Ecolanda's 7-generation stewardship. Abundance methods avoid extraction, which ensures that plentiful natural resources will be available for future generations to enjoy.

Regenerative technologies are highly productive as they employ microcrops to produce healthy and clean food, feed and other bioproducts. The abundance cultivation process cleans, repairs and restores ecosystems degraded or destroyed by mines, refineries and industrial production.

Ecolanda's regenerative technologies begin with biocycling carbon and other pollutants from the atmosphere. Every ton of microcrop food, feed or fertilizer captures two tons of CO_2. Microcrop cultivation cleans and clears the atmosphere we share. Polluted air bubbles through biosystems where tiny microcrop cells removes contaminants. Extreme contamination may require more than one pass through a biosystem.

Ecolanda's regenerative biosystems can clean polluted water and restore extra fresh blue water for drinking and other community needs.

Industrial production has seriously degraded verdant ecosystems and destroyed their colourful biodiversity. Ecolanda's highly productive smart biosystems repair and restore health to degraded forests, fields and waterways.

Healthy ecosystems are attractive to birds, bees, butterflies and other land and marine creatures. Restored ecosystems allow bright and beautiful biodiversity to repopulate and sustain biodiverse life for many generations.

Ecolanda priorities

Ecolanda growers orchestrate smart biosystems to optimize health, productivity and safety for people, producers and our planet.

Industrial farmers suffer among the highest health, disability and premature death danger of any occupation. Farmers are injured by huge equipment and physical labor and poisoned by dust, black soot, fertilizers and pesticides.

Biosystems assure grower safety. Cultivation requires no heavy physical labor or large tractors or harvesters. Growers are not overwhelmed by dust and are not exposed to hazardous chemical fertilizers or agri-poisons.

Even if technology advances removed all air, water and ecosystem pollution, industrial systems are so inefficient they create far too much waste.

Waste adds expense to every step in the industrial supply chain. Industrial agriculture typically wastes more than 50% of all the inputs farmers apply to their crops. Those wastes massively pollute our air, water, ecosystems and oceans.

Ecolanda operates efficiently across five highly integrated vertical sectors: agri, energy, waste, water and a smart ecocity.

Ecolanda will create the **highest global:**
- Sustainable food yield per hectare
- Nutrient density and diversity per bite
- CCU, carbon capture and utilization, reuse
- Sustainability metrics & smart biosolutions
- Renewable energy generation, distribution, storage and use efficiencies

Goals include the **lowest global**:
- Carbon, water and ecological footprint
- Net food and bioproduct production cost
- Natural resource extraction, waste, emissions and pollution
- Cropland, fossil fuel and water use
- Chemical fertilizer and pesticide use

A National Task Force in each host country selects additional National priorities.

National priorities

The SCAD Ecolanda team co-plans each Ecolanda project priorities with the National Task Force. Most countries have National Priorities for health, economy, society and environment.

The two teams jointly create a set of goals, timelines and accountabilities in each area. Example goals are:

Environment
- Stop biodiversity loss
- Lead the world in decarbonization
- Slow deforestation by 80%
- Reclaim 20,000 hectors of degraded forest
- Demonstrate cleantech biosystems that pay for themselves while they repair ecosystems

Health
- Reduce malnutrition 80%.
- Reduce preterm newborn deliveries 60%
- Reduce stunting and wasting 50%.
- Improve health and vitality 30%.

Economy
- Increase employment 30%.
- Improve rural economies 20%.
- Increase exports 20%.
- Upgrade clean-tech innovation and entrepreneurship 40%.

Society
- Increase life-quality and happiness 30%.
- Enhance key education metrics 50%.
- Improve social satisfaction 30%.
- Improve community engagement 50%.

National goals change with experience and demonstration of successful novel biosolutions. The SCAD Ecolanda team and the National Team will prepare a comprehensive National Goal progress report annually.

Models and projections play an important role in Ecolanda metrics. Data on goal progress allow the National Team and institutional experts to model smart systems. These models help the National Team make policy decisions the benefit communities and the nation.

Respect

SCAD's primary purpose focuses on lifting life quality for people, producers and ecosystems.

SCAD respects People and ensures that People there are healthier, happier, better educated and better employed then when SCAD arrived.

Transparency

In many areas of human endeavor, especially health, economics, society and environment, superior decisions depend on accurate metrics.

Ecolanda success depends on robust metrics. Mechanical industry uses general measures to assure industrial production stays within certain parameters.

Biosystems are both more complex than mechanical methods and much smaller. Nano and microorganisms are usually not visible without a microscope. SCAD's smart systems use 5 to 10 times more metrics than their mechanical analogs.

Ecolanda's smart systems operate with sensors and monitors that provide big data 24/7. All inputs and outputs are measured in order to optimize operations and scout for improvements.

Live metrics flow to dashboards that assemble the data in templates that are easy to read. Key people on the National Team or their designees have access to live dashboards.

Selected experts at National or State Academies and Universities will have access to these reports. The reports and data behind the reports will allow both internal and external experts to analyze and validate progress.

Dashboards roll up into daily, weekly and monthly reports. These rollups provide quick biosystem efficiency summaries. Reports help spot inefficiencies or choke points when they occur.

SCAD has created strong alliances with world-class scientists globally that will have access to selected data and reports. These subject matter experts will provide third-party validation for progress on goals and objectives in each area.

Critical export

The most important Ecolanda export will be circular econometrics. The metrics compare traditional industrial production with biosystems.

The first question most people ask is: "If biosystems are superior in so many ways to mechanical systems, why do investors continue to select industrial methods?"

The metrics for proof do not exist today. No large integrated circular biosystems currently exist. Investors prefer to put their money in models that generate an immediate and known return. Investments continue even if the processes are viable only short term and continuously impose havoc on our environment.

Ecolanda will make the critical eco-metrics available free globally. Metrics will allow investors clean options when deciding where to place their money.

The second argument against clean investment is: "It costs too much." Eco-metrics will show clearly that communities and nations are paying a far higher price today for dirty industrial production. They can shift their existing investments to cleantech and save money and their environment while creating great new jobs.

Site selection

SCAD uses a comprehensive matrix for site selection. Selection begins with a site visit to study the tentative location and determine if the area fits Ecolanda development parameters. Site selection examines factors supporting Ecolanda's integrated systems model. Site selection allows SCAD leadership to determine the most appropriate location.

Finance

Ecolanda becomes a project of **"national interest"** due to its importance and social and economic scope. It is expected to be the largest private investment in a single project in the country's history. Ecolanda will bring economic gains of more than €200 billion within ten years.

Project financing varies by country. Ecolanda projects are so large that national leaders need to be highly engaged in the process. Extensive coordination between national leadership and SCAD are critical to build Ecolanda successfully.

The public part of the public-private partnership begins with National Leaders making several commitments:

1. Appoint a strong interdisciplinary National Task Force to coordinate Ecolanda designs, plans, budgets and timelines with SCAD.
2. Issue a government bond for €1 – 2 billion (Based on Project Scale) to a top 10 bank.
3. Designate minimum two land parcels of 10,000+ hectares, each for a 99-year lease.
4. Agree to appropriate in-kind infrastructure improvements and tax incentives
5. SCAD site to be designated an Economic Trade Free Zone for 30 years sustaining the legal conditions for investment.
6. The government's decisions contribute to project competitiveness and reduces construction costs.

The Economic Trade Free Zone will:
- Boost exports and foreign exchange.
- Create jobs, training and a wide array of advanced technologies.

SCAD orchestrates the private sector part of the partnership with aligned actions:

1. A SCAD interdisciplinary task force will coordinate with national, regional and local entities via a National Task Force.
2. SCAD will commit to a €15+ billion private investment to build Ecolanda in the country.
3. Present architectural designs, plans, timelines and accountabilities.
4. Share regular progress reports on a mutually approved schedule.

SCAD will communicate with National Task Force on investment timing. SCAD creates a broad project strategy with specific plans. Timelines on plans may change based on mutual agreement among the relevant parties.

The next section highlights how Ecolanda uses Abundance agricultural methods to accomplish both high productivity and ecological goals.

9. Ecolanda

Vision: Ecolanda leads the global transformation to clean agri-energy with zero waste or pollution and restores health to consumers, producers, and our environment.

Societies globally are systematically running out of natural resources. Scarcity constantly drives-up manufacturing cost and consumer prices. Every minute, industrial production extracts more irreplaceable natural resources to feed its insatiable appetite.

Industrial systems accelerate fossil resource loss, overconsumption, waste and pollution. Cropland has become so scarce that farmers are clear-cutting ancient forests to grow animal feed.

All food-growing continents face water scarcity. Entire cropland regions are abandoned as wells and surface water sources go dry. Fossil fuels continually pollute our air with CO_2, poisonous nitric oxides and deadly black soot particulates.

Chemical fertilisers erode from cropland and have created over 500 dead zones globally in waterways and oceans where eutrophication has suffocated all life. Deadly pesticides kill good and bad insects and exterminate biodiversity.

When our children need natural resources, they will be gone. Industrial emissions continually accelerate global climate chaos. Mechanical systems systemically degrade and destroy ecosystems and their colourful biodiversity.

Population growth drives industrial production in the wrong direction, leading to more overconsumption, GHG emissions and ecological contamination. Intensive mechanical agriculture, IMA, destroys our ecosystems and biodiversity.

Our children deserve a change in direction.

Ecolanda Strategy: Abundance

The world needs a new paradigm that produces superior food and other consumer goods faster and sustainably. New methods should assist natural systems with accelerated environmental restoration and biodiversity recovery.

Ecolanda addresses each of these critical issues by transforming industrial mechanical production to biological solutions using **abundance** methods.

Ecolanda transforms industrial food and consumer products creation to abundance methods that avoid natural resource extraction, overconsumption, waste and pollution. Abundance cultivation may occur on non-cropland. This saves valuable ecosystems and their biodiversity for future generations.

Circular Bioeconomy

Abundance methods enable the creation of a 360bioeconomy. Circular biosystems provide the most sustainable and ecologically responsible large-scale development. Biosystems maximize health, economic and social goals while repairing and restoring the environment.

Ecolanda uses the circular BioRenew process to enhance nutrition in a manner that delivers healthier food for people, animals and plants.

Decarbonize

Removing carbon from the atmosphere improves air quality by eliminating GHG and other toxic chemicals from industrial stacks. Improved air quality improves health, reduces respiratory diseases and enhances life quality.

Most communities need more blue drinking water. The BioRenew process creates millions of liters of blue fresh water annually from agricultural, industrial and residential waste streams. Ecolanda's goal: create 20% more fresh blue water than the project uses.

Nrich restores nutrients and fertility to damaged and degraded croplands. Soils often suffer from nutrient depletion from years of farming. Repeated crop production also extracts micronutrients and humus, the critical organic matter that holds moisture and nutrients.

Nrich can restore macro and micronutrients as well as humus to cropland. Nrich can clean and restore ecosystems. These integrated systems allow biodiversity to recover.

Ecolanda creates over 10,000 direct jobs and more than 50,000 indirect jobs. These jobs provide sustainable, long-term benefits to society and lift the economy and living standards.

Ecolanda cultivates affordable food, feed, fibre and a wide array of other valuable bioproducts. The agri-energy sector produces health and affordable food for over 5 million people.

Sustainable food cultivation ensures continuity and improved quality for the region and for the nation's food supply. Ecolanda also cultivates food, feed, fibre and renewable energy for local markets and for export.

Ecolanda strategies

The Ecolanda architecture applies abundance strategies to address substantial economic and ecological challenges. These approaches provide a 180° disruptive change from modern industrial mechanical production.

Ecolanda strategies apply Abundance methods and integrate across sectors to:

Ecolanda — BLINC Strategic Differences	
Industrial Production	**Abundance Methods**
1. Mechanical devices	1. Biological systems
2. Fossil energy	2. Light energy
3. Independent systems	3. Integrated systems
4. Land-based crops	4. Nano and microcrops
5. Linear economy	5. Circular bioeconomy

1. **Stop systemic damage** inflicted by industrial methods – extraction, overconsumption. waste, emissions and pollution.
2. **Repair ecosystem damage** caused by destructive industrial processes and allows the restoration of biodiversity.

Industrial agriculture offers sharp contrast to abundance methods. Industrial manufacturing does similar damage for other consumer products.

Mechanical agriculture

Industrial mechanical agriculture drives huge fossil-fuel machinery designed to strip broad expanses of natural ecosystems. Machinery consumes massive fossil resources in order to force nature to do the farmer's will.

Mechanical methods systematically abuse cropland with tractors, ploughs, disks, rippers, compactors, harvesters, fertilisers, pesticides and erosion until the living soil dies from exhaustion.

After years of abusing soils, farmers abandon destroyed cropland. They leave millions of once pristine cropland hectares degraded, destroyed and unavailable for future generations. More farmland has been abandoned globally in the last 60 years than is farmed today.[128]

Save Biodiversity

Intensive mechanical agriculture

Farmers employ heavy mechanical equipment to:

1. Carve out nature - scrape the natural ecosystem "clean" to make way for GMO monocultures.
2. Slash, gash and rip the topsoil deeply to prepare the soil bed.
3. Destroy biodiversity.
4. Disk the topsoil flat to prepare for planting.
5. Crush and compact the soil with monstrous heavy equipment.
6. Degrade living soil with herbicides and cultivation.
7. Crush soils with irrigation and wet-ground compaction.
8. Kill microorganisms with tons of fertiliser, herbicides, fungicides and pesticides.
9. Erode topsoil, fertilisers and pesticides that migrate into waterways and rural ecosystems.
10. Harvest once a year.
11. Extract huge stores of nutrients and humus with every crop, without replacement.

Farmers produce fossil-based foods that suffer from nutrient dilution and hidden hunger. Crops cannot pull nutrients from deficient soil. Nutrient-deficient produce transfer their micronutrient deficiencies to our children and to all consumers.

Biological – Abundance methods

Biological systems extract nothing. They do not overconsume natural resources or create waste. Nature's integrated sustainable systems do not pollute air, water or soil. Nature's circular system works exquisitely with the cycle of life.

Nature nurtures food cultures gently without mechanical intrusion. Nature has used biological systems to grow nutritious food for trillions of consumers daily for several billion years.

Ecolanda growers use abundance biological methods that work in harmony with nature to cultivate healthier, truly sustainable freedom foods with microcrops grown with free energy – sunshine.

Abundance methods allow nature to work gently, yet highly effectively. Microcrops grow 20 - 50 times faster than fossil-based IMA crops.

1. Build a raceway or indoor biosystem and cultivate microcrops.
2. Energy – gravity and sunshine. No fossil fuels.
3. Zero fresh water, brine, waste or ocean water.
4. Zero chemical fertilisers
5. Zero pesticides, herbicides or fungicides.
6. Restores ecosystems – air, water, and biodiversity, flora and fauna.
7. Harvest daily or every few days, all year round.

Abundance growers cultivate freedom foods that are healthier for people, producers, animals and our planet. Freedom foods may be made in any form, shape, taste or colour.

Nature continually cleans and restores health to air, water, soil and ecosystems. Natural systems require hundreds of years to repair damage from only a few decades of industrial production.

Ecolanda offers several paths to accelerate nature's ability to repair ecosystems and restore biodiversity with novel biotechnologies.

Light energy

Abundance growers use light, solar energy, and photosynthesis as the engine of choice in place of fossil fuels to biocycle nutrients. Light energy may be augmented with concentrated light forms such as photovoltaic and concentrated solar.

Renewable solar and wind energy drive integrated processes across the five smart sectors.

Instead of extracting mined chemical fertilisers, growers use the natural photosynthetic-driven BioRenew process to capture, recycle and repurpose carbon and other nutrients from waste streams.

Reusing nutrients efficiently saves growers money and avoids massive emissions and pollution.

Ecolanda farmers cultivate food in sustainable bio-systems that mitigate climate change as they assimilate rather than emit carbon.

The only emission from a biosystems is pure oxygen. Biosystems enhance social and economic growth free of resource over-consumption and pollution.

Abundance methods flip food cultivation from a major GHG emitter, about 36% of global emissions, to a negative carbon process. Each Ecolanda project plans to capture over 70 million metric tons of CO_2 equivalents annually.

Integrated systems

Ecolanda differs from other agri and economic development because the architecture mimics nature. Each area – agri, energy, waste, water and the ecocity – integrate in a symbiotic manner. The natural architecture allows substantial resource savings compared with IMA and other industrial development processes.

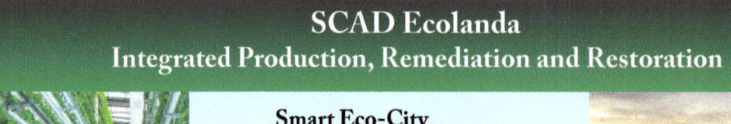

SCAD Ecolanda
Integrated Production, Remediation and Restoration

Smart Eco-City
- **80% water savings**
- 70% energy savings
- Gardens, green space
- Healthy lifestyles
- High connectivity
- Green transport

Smart Energy
- **100% less fossil fuel**
- Solar, wind, hydro
- Geothermal, tides
- Smart microgrids
- Advanced storage

Smart Agriculture
- **Biocycle carbon**
- Higher productivity
- Superior nutrition
- Healthier food
- Abundance methods

Ecolanda
- Superior health
- Higher productivity
- Zero waste
- Zero pollution

Freedom foods are free from fossil resource consumption, waste and pollution.

Smart Water
- **20% more blue water**
- Brine and waste
- Ocean and brackish
- Clean & monitored
- Minimize water loss

Smart Waste
- **100% less landfill**
- Zero GHG
- Near-zero runoff
- Waste to energy
- Waste to bioproducts

Sustainable agri-energy requires extensive energy, water and waste management. SCAD applies a suite of renewable biological and energy technologies to sustainably manage these critical resources.

Traditional agriculture and animal production are well-known massive emitters of severe pollutants to air, water and soils. SCAD technologies mitigate those pollutants and pursue a goal of net-zero air, water and soil pollution. Ecolanda ensures that no wastes are burned in the open air or buried in waste dumps.

Mechanical agri-energy methods act alone and each waste and emit their own carbon and other pollutants. Each Ecolanda abundance system connects with others to ensure waste from each activity is captured and reused by paired processes.

Smart energy

Smart energy supplies each sector with renewable energy. Agri-systems both use and produce waste-to-energy and biofuels. Waste-to-energy systems create additional energy in the form of syngas, green hydrogen and other energy products for local use and for export. The smart ecocity produces over 80% of its own energy and food in the municipality.

Smart water integrates across sectors. Industrial agri-systems use massive amounts of blue water. Salt in soil or brine water kills land-based plants due to a plumbing problem. The large salt ions clog the roots, which stops nutrient circulation and the plant withers and dies from lack of nutrients.

Microcrops have no roots and flourish in brine or other non-potable water, including sea water. Most communities have large amounts of wastewater. Smart water treatment and microcrop biosystems allow continual water treatment. Water may be reused many times.

Nano and microcrops

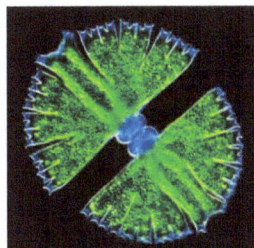

Nano and microcrops are tiny, often just five microns small. These single and multicellular plants without roots grow in all types of water.

Algae are small but powerful. Algae produce 70% of the new oxygen on earth daily, more than twice the oxygen from all the forests and fields combined.

Algae produce 40% of the new biomass on the planet daily. Trillions of consumers eat algae because algae are the most nutritious food.

Tiny microcrops have a huge surface area. A cube of algae the size of a postage stamp, (2 cm x 3 cm) contains more surface area than three football fields. Each algae cell operates independently. The large surface area allows extremely fast assimilation of nutrients in each cell.

Algae grow biomass 20 to 50 times faster than terrestrial crops such as rice or corn. A healthy nano-culture can double its biomass each day.

Ecolanda's microcrops are nurtured gently without large mechanical tractors or trucks. Some biosystems use paddle wheels while others mix cultures by bubbling in air and CO_2. Growing microcrops does not require ripping the land with ploughs and disks. Growers do not have to clear the land and destroy biodiversity.

Industrial farmers often drive big heavy tractors across their cropland seven or more times a season cultivating, thinning, tending to weed invasion, fertilization, applying pesticides and herbicides and harvesting. Nanocrops require none of these mechanical processes, which saves considerable energy, pollution and costs.

A corn or maize farmer uses mechanical devices to add expensive inputs to the crop. This requires a full growing season, 120 days, to produce the first gram of food. A 4-meter maize stalk supports a big biomass but contains less than 2.5% food. Only the kernels on the single cobb are edible and contain just 33% of an incomplete protein.

Industrial farmers must pay for all the inputs to produce enormous waste biomass and then pay again for waste disposal. Most economists would question the sensibility of producing food if 97% of the biomass produces only waste.

Microcrops are harvested daily or several times a week, year-round. A microcrop may contain 65% edible protein in every kilo of biomass. The residual biomass contains additional edible and valuable components – vitamins, minerals, oils, lipids, carbohydrates and other compounds. Microcrops may contain 9% ash, which also contains residual nutrients. Ash residues are biocycled in subsequent cultures.

Ecolanda's freshwater savings, 100% compared with industrial farms, comes from growing microcrops for food, feed, biofertiliser, medicines and other bioproducts in non-potable water.

Ecolanda growers cultivate an extensive array of microcrops like algae with net-zero blue water.

Microcrops cultivated in outdoor raceways

Water substitution saves blue drinking water. Blue water contains less than 1,000 parts per million, (ppm) saline. Brine salinity, (NaCl) measures higher than blue water and can be 10 times saltier than ocean water, which contains 36,000 ppm salt.

Most communities can find brine water available since half the water stored in the earth is brine.[129] In many places, brine aquifers are so close to surface, water can be removed by a foot pump.

A nutraceutical project in Arizona uses brine water in vertical column photobioreactors to cultivate algae with omega-3 fatty acids, (below).

The project pumps brine water from an aquifer under the Painted Desert left by an ancient ocean.

The highly saline water is used for nine algae cultivation cycles until algae removes most the nutrients. After nutrient extraction, the still salty water flows to a salt lagoon where the water evaporates, and the salt recovered.

Linear industrial production

Industrial firms today follow an inefficient linear production model that imposes high operational costs, relies on increasingly expensive fossil resources and levies an appalling toll on human and producer health and the environment.

Intensive mechanical agriculture has provided food and other goods for the recent 70 years. Modern consumer products come at the extreme cost of extracting and over-consuming natural resources that will be gone when they are needed by our children and their children.

Mechanical methods require huge amounts of fossil energy because fossil fuels power tractors, trucks, manufacturing and other systems.

Wasteful industrial methods systemically make our planet less livable. Industrial waste pollutes and degrades our air with GHG, smog and black soot particulates.

Mechanical agriculture imposes extreme waste to produce low energetic crops like maize where the waste biomass to food ratio is 97% waste, 3% food. Half the agri-inputs used to grow crops go to waste too, including freshwater, fertilizer and pesticides. These wastes pack waterways with eroded toxic chemicals and deadly pesticides.

Croplands have been severely eroded and exhausted. Large cropland regions have been abandoned because soils have been stripped of nutrients or wells have gone dry.

Linear production causes these vital resources to be extracted and transported to farms where they are stored and then used very inefficiently only once. Each year, farmers must repeat the linear process of buy, store, apply to fields and reorder for next year. Farmers have little control over the substantial cost of waste because field crops use inputs inefficiently. Nearly all industrial agricultural input costs are rising due to increasing scarcity.

Waste, which often exceed the weight of the food crop, are often burned or buried. Crop input wastes are carried away in eroding soils, creating extensive air, water and land pollution.

Save Biodiversity

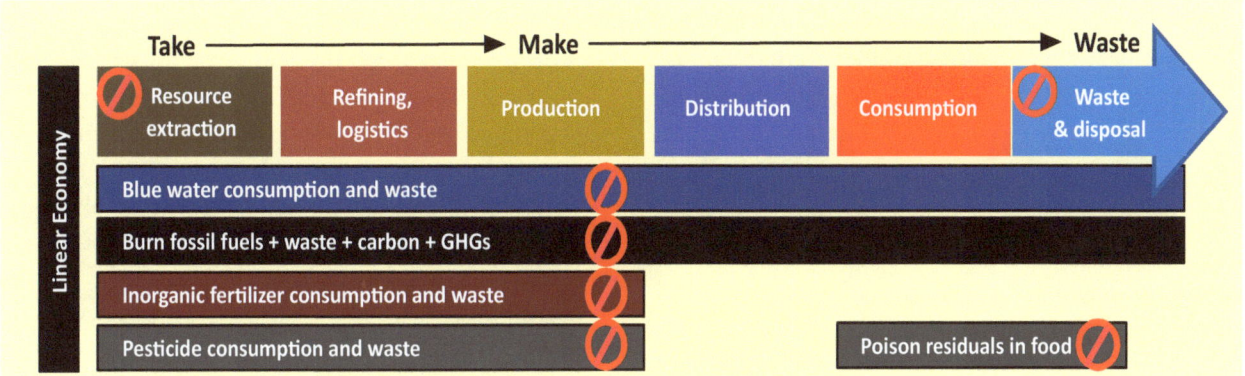

The linear economy model illustrates the extreme waste caused by industrial production. Modern agricultural wastes – cropland, fresh water, fossil fuels, inorganic fertilisers, pesticides – pollute, degrade and poison air, rivers, lakes and oceans.

Crop residues, animal dung and other wastes are often burned creating huge toxic plumes filled with GHG, smog, nitric and sulfur oxides and deadly black soot particulates.

Poor air quality results in early onset of respiratory diseases such as asthma. Prolonged exposure results in respiratory failure and premature death. Oceans are polluted with so much industrial waste that many coastal communities cannot farm fish in the sea. Some seas have more plastic than fish.

A comprehensive study, *The Water Footprint of Humanity*, found that industrial agriculture consumes an incredible 92% of available fresh water.[130] Surface irrigation loses over 50% of the blue water to evaporation before it reaches crops.

Less than half of the chemical fertilisers applied to croplands are assimilated by the plants. The remainder erode on wind and rain and invade local ecosystems. Less than 1% of pesticides are assimilated by the crop, leaving 99% to invade and poison local waterways and groundwater.

Numerous scientific studies have shown that dangerous poison residuals remain on produce when they find their way into consumers' homes.[131]

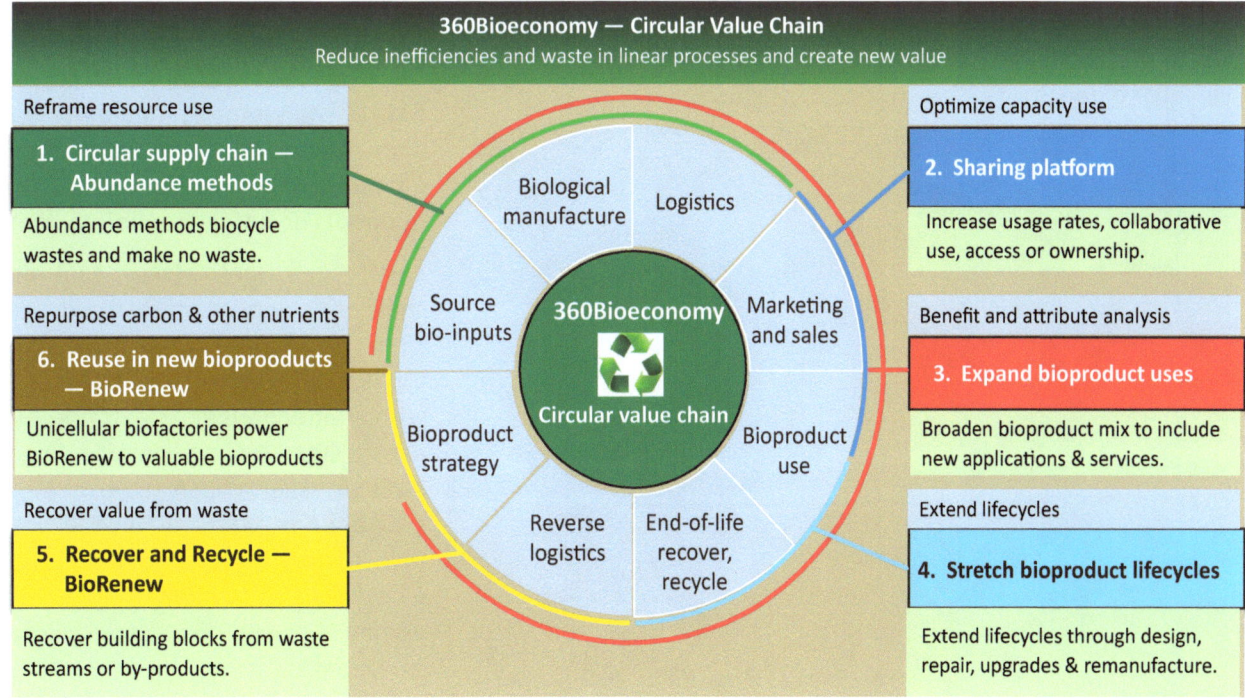

Mark R. Edwards

Ecolanda circular value chain

The significant difference between Ecolanda and other economic development is SCAD's advanced architecture for a circular bioeconomy.

Ecolanda uses a circular value chain that overcomes linear economy problems. SCAD avoids natural resource extraction with BioRenew, which biocycles carbon and other nutrients from waste streams.

The inner circle in the circular diagram shows SCAD's key organizing divisions for the circular value chain. These units integrate across agri-energy-waste-water and smart ecocity.

A 360Bioeconomy represents a sharing platform for biomanufacture. Collaborative use of systems and technologies improves system integration and process efficiencies.

Industrial manufacturing typically creates excessive waste due to the design for a single product. The bioeconomy model changes to expand bioproduct uses and users. Many bioproduct users are local, which reduces costs.

Essentially any product made with industrial manufacture can be made with the SCAD 360Bioeconomy. The ecocity innovation park prioritizes biomanufacture of consumer and industrial products based on social and economic factors. The graphic illustrates the wide array of valuable bioproducts.

Bioproducts

Bioproducts share important attributes. They are all made with biological processes, so they are biodegradable. Microcrop construction materials supply equal or superior features to industrial versions. When bioconstruction materials come to the end of their useful life, possibly 50 years, they are recovered and recycled. Target compounds are repurposed in new products.

Bioproducts can be made using zero fossil resources, which makes them less expensive and eco-friendly. Many bioproducts such as asphalt and cement substitutes sequester vast amounts of carbon and other polluting substances for generations.

Bioplastics, films, packaging and foams provide valuable substitutes for industrial products that currently are sent to waste dumps and flow into waterways and oceans. Bioproducts break down quickly and do not create land or ocean litter.

A maize farmer grows and sells only maize, the tiny kernels on the cobb as a single commodity product. Microcrops are cultivated with a plan for multiple products from the single culture. Growers tend to avoid commodity products because specialty bioproducts bring more value.

Processing algae biomass may remove 20% of the oil for an Omega-3 fatty acid and the remainder of the oil for cooking. The protein, which may be 65% of the biomass, may be sold as animal feed. The carbohydrates and fibres may be used for construction materials or ruffage for animal feed.

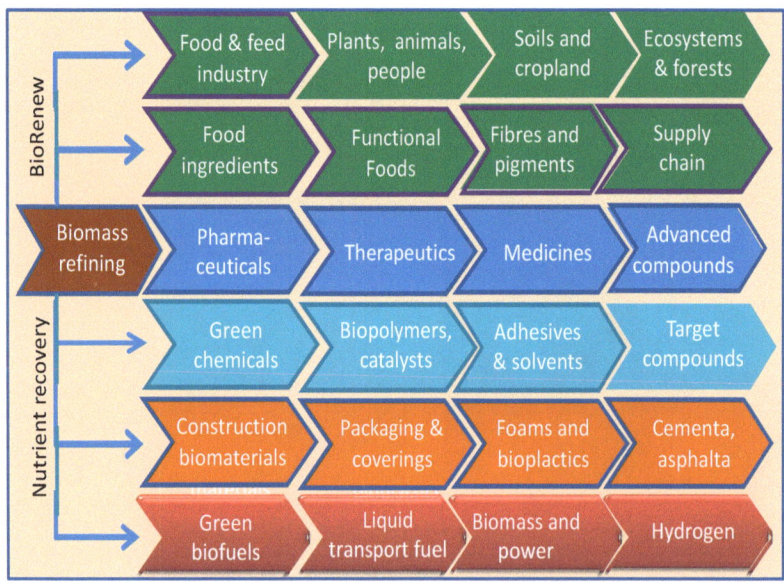

Fossil fuels, petroleum, coal and shale are compsed of fossilized algae from ancient oceans. Nature fossilized the algae deep in the earth with tremendous heat and pressure over 400 million years. Algae biofuels skip those steps. Algae growers can cultivate jet-fuel in a few weeks and produce more fuel weekly. Algae biofuels burn cleanly without smoke or black carbon particulates.

10. Nrich – Superior Nutrition

> Plants, animals and humans
> are what we eat. – **Lucretius**

Fossil fuel powers mechanical engines. Richer fuel improves engine performance.

Nutrition powers cellular metabolism; the bioengines for plants, animals and people. Nrich bioenergy improves metabolism, health and vitality. Algae biomass and foods deliver superior nutritional energy for plants, animals and people.

Nrich biofertiliser enhances crop growth by >20%. Other attributes improve too such as quality, yields and taste.

Nrich biofeed enhances animal growth by >10%. Fish grow faster, excrete less waste and benefit from stronger stress tolerance.

Nrich creates plant-based meat with 2.5 times the protein per bite as beef. The food is healthier for people and our planet. Animals are said to prefer non-meat alternatives too.

Algae, the first food on the planet offers healthier nutrition – lipids, proteins and carbohydrates. Lipids are long carbon chain molecules. Lipids store energy for the plant and serve as the structural components of cell.

Proteins are large organic compounds made of amino acids arranged in a linear chain connected by peptide bonds. The plant's genetic code determines the sequence of the amino acids, but nutrient limitations may cause changes to the production of amino acids.

Most proteins are enzymes that catalyze biochemical reactions and plant metabolism. Other proteins maintain cell shape and provide signaling functions. Algae deliver more and healthier proteins per bite than any animal meat or any land plant.

Quality animal feed demands protein. Professor E.W. Becker wrote an excellent review article on algae animal feed. He concluded the quality of algae protein is equal, and in many cases superior to conventional plant proteins, including soy.[132]

Starches are complex carbohydrates which are insoluble in water. Plants use starches to store glucose as plant sugar.

Biomass composition among algae species varies tremendously. Some algae hold 80% lipids while others are 65% protein. Some macroalgae, seaweeds contain 92% carbohydrates. Species selection is critical, not just for the desired composition, but for a host of micronutrients and growth biostimulants that vary across cultivars.

Algae varieties offer almost infinite nutrition combinations and useful bioactive compounds. Special attributes such as omega-3 oil production can be enhanced through selection screens for naturally occurring organisms and mutagenesis, which is similar to a rapid hybridization process.

Some companies hybridize algae to express more, or less of target compounds. Algae biofeed will become far more desirable as specialty compounds that help animals with digestion, biosorption and protection against pathogens are discovered and used in algae cultivars.

Algae biofeed provides the full set of essential nutrients for animals. Considerable research focuses on protein quality, which varies across various species of terrestrial plants and algae. Unsurprisingly, different types of animals grow faster with high quality proteins combined with a full set of micronutrients.

Algae composition displays considerable variation with seasons. Algae biofeed formulation requires continuous monitoring, similar to field grains. Traditional crops such as maize are often mixed with antibiotics to improve digestion. Omega-3 fatty acids may be added to improve animal health.

Algae biofeed can eliminate pharmaceuticals as feed additives because digestion does not cause problems. Various algae species provide different amounts of omegas and other specialty oils.

No terrestrial crop or animal produces omega-3 fatty acids naturally. Some have been genetically modified to produce a low-quality omega antioxidant, but research has not established whether the molecule is bioavailable to animals.

Algae biofeed can deliver the omegas of choice.

1. **Linoleic acid, LA,** is a non-essential unsaturated omega-6 in soaps, emulsifiers, quick-drying oils and beauty aids.
2. **Arachidonic acid**, **AA**, is non-essential and may induce inflammation.
3. **Eicosapentaenoic acid, EPA,** an omega-3 fatty acid that gives the same benefits as fish oil.
4. **Docosahexaenoic acid, DHA,** an omega-3 fatty acid is the most abundant fatty acid found in the brain and retina. DHA deficiencies cause cognitive decline and higher brain cell death.

Many algae species are tolerant of wide variations in growing conditions. Some species are nearly blind to geography. All algae species can be grown in CEA vertical farms.

Algae biofeed nutrition

The nutritional contents of algae are superior to land-based animal feed. Algae biofeeds provide a renewable substitute for conventional animal feed ingredients for meat, dairy, poultry and fish.

SCAD Biofeed Animal Nutrition and Productivity	
Animal nutrition:	**Animal productivity:**
• Stronger vitality	• Faster growth, 10%
• Higher nutralence, 30%	• Higher survivability, 10%
• Higher palatability, 20%	• Stronger vitality, 20%
• Bioavailable nutrients, 40%	• Higher stress tolerance, 20%
• Less waste, 10%	• More robust health, 20%

Algae contain all the vitamins, minerals and micronutrients needed for healthy biofeeds, including Vitamins A, B1, B2, B6, B complex, C, D, and E. Algae are rich in niacin, iodine, potassium, iron, magnesium and calcium.

SCAD Algae Biofeed Compared with Field Grains	
Higher productivity:	**Ecological savings:**
• Growth rates, 20 to 100%	• Ecological efficiency, 80%
• Productivity / ha, 50 to 100%	• Less waste, 80%
• Protein / ha, 50 to 100%	• Water efficiency, 50 to 80%
• Omega-3s, 100%	• Diesel fuel efficiency, 90%
• Quality, 20%	• Inorganic fertilizer, 80%

Algae biofeed grows with substantially higher productivity per hectare than grains or palm oil. Each algal cell contains polysaccharides, sugar and starch, as well as iron, sodium, phosphorus, magnesium, copper and calcium. Algae biofeeds deliver nutritional benefits not available in land plants, antioxidants and bioactive compounds.

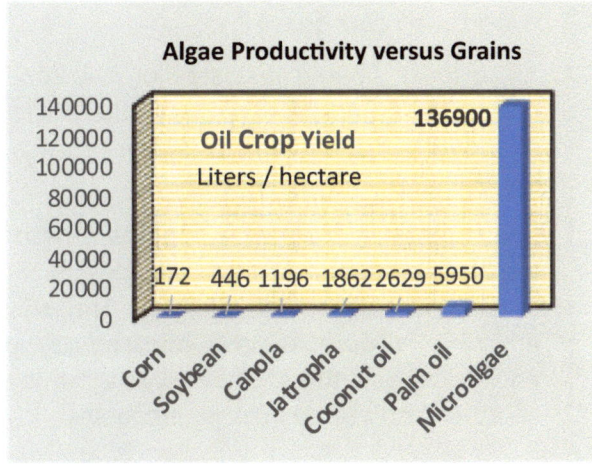

Algae productivity for protein per hectare is 2.5 times higher than oil. Algae biofeeds provide a rich source of high-quality protein, vitamins, micronutrients, trace elements and carotenoids.

Algae produce valuable biomolecules including astaxanthin, lutein, beta carotene, chlorophyll, beta-1,3-glucan and phycobiliprotein.[133]

Many valuable biomolecules in algae are not synthesized in the animal, (or human) body. However, they are considered essential for healthy body, brain and heart growth and development. Therefore, animals must get them through their diet.

Algae provide both an ideal nutrient package and delivery system. The algae package, a cell, is so small it becomes immediately bioavailable to the animal. The algae delivery system provides higher nutralence than conventional animal feed, with significantly higher nutrient quality and density,

Nrich biofertiliser

SCAD Biofertiliser Compared with Industrial Fertiliser	
Improves crop:	**Improves market value:**
• Germination rate, 20%	• Taste and aroma, 20%
• Time to maturity, 20%	• Vitamins & minerals, 50-100%
• Health and vitality, 30%	• Digestible nutrients, 50-100%
• Yield and quality, 30%	• Color and texture, 20%
• Produce size, 20%	• Shelf-life, 25%

Algae biofertiliser significantly improves seed germination, early growth, speed of growth to maturity and survivability by more than 20%.

Multiple projects globally have demonstrated biofertiliser improves produce yields in excess of 20%. In addition, produce display superior nutrition that avoids hidden hunger and micronutrient deficiencies. Improvements in produce market value arise from the unique Nrich nutrient delivery system that promotes faster growth to maturity and harvest.

Most modern farms constantly extract nutrients but replace only N, P, K macronutrients. Nrich biofertiliser delivers the full set of 43 macro and micronutrients plants need for successful growth. Nrich also delivers key vitamins, minerals and trace elements, which improve crop stress tolerance as well as colour, taste and texture.

SCAD Biofertiliser Compared with Industrial Fertiliser	
Reduces production costs:	**Enhances soil:**
• Tillage, 30%	• Porosity, looseness, 500%
• Diesel fuel, 40%	• Microbial life, 500%
• Irrigation water, 25%	• Erosion resistance, 30%
• Inorganic fertilizer, 80%	• Bioavailable nutrients, 50%
• Pesticide / herbicide, 90%	• Organic material, 20% / year

Nrich relies on BioRenew for nutrient capture, recycle and repurpose. Larger produce with a longer shelf life underscores the value of the full nutrient set delivered by algae biofertiliser. Only healthy produce can resist spoilage for several extra days. Nutritional and productivity increases improve grower yields and profitability. Ecolanda provides a plentiful continuous cycle of new waste stream nutrients.

Nutrients may be captured from air, water and biosolids including animal manure or coal dust. These nutrient waste streams serve as building blocks for new green biomass that may be transformed to biofertiliser, feed or a wide array of valuable bioproducts.

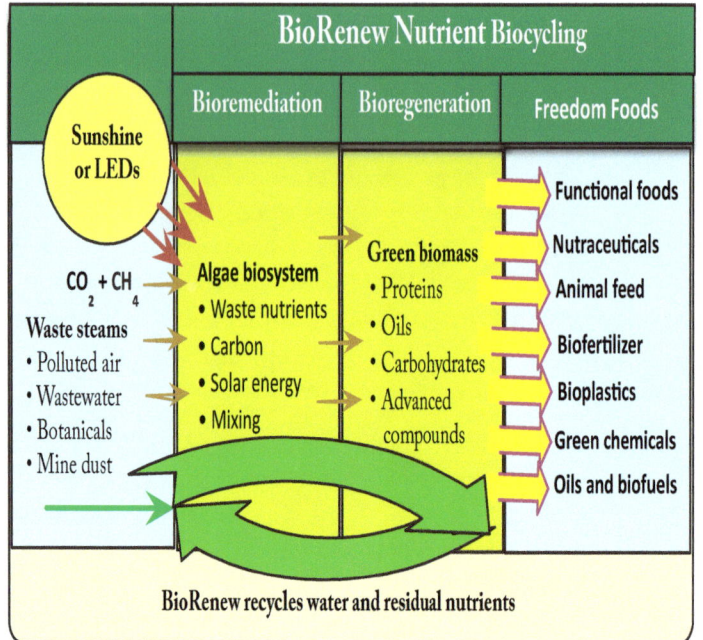

Growers may grow the same crop for months or change the culture every month or every season to a different algae species that provides more of the desired target compounds.

Microcrop diversity

Scientists estimate there are about 300,000 land-based plants. Only 17 low-nutrient plant species are consumed as 90% of the global human diet.[134] Nature has provided over 10 million species of algae.[135] Each produces its unique set of target compounds.

Most algae species produce significantly higher nutralence than industrial field crops.

This diversity gives growers a wide choice of cultivars. Most growers today cultivate one or more of 30 algae species. Each cultivar may offer higher or lower productivity and target compounds based on growing conditions. Improved cultivation models will increase species diversity and productivity.

Ecolanda growers avoid the use of pesticides and poisons. Algae biofertiliser provides a wide array of plant hormones and biopesticides that protect the plants. Avoidance of pesticides assure that no poisons enter the local waterways or tag along as residuals on consumer produce.

Algae use solar energy

Algae make up the first step on the food chain. Each algae cell contains all the crucial elements needed for healthy growth, development and reproduction.

The mechanism for action, photosynthesis, transforms water and carbon dioxide to plant sugars in a carbon-rich biomass. The only emission is pure oxygen released in in enormous quantities. The oxygen can be captured and reused in other processes.

Photosynthesis uses free sunshine for energy. Energetic photons streaming with sunshine drive the process. Algae and other plants are agnostic regarding the photon source. Growers may supplement sunshine from other indirect solar energy sources such as fibre optics, lasers, mirrors or LED lights.

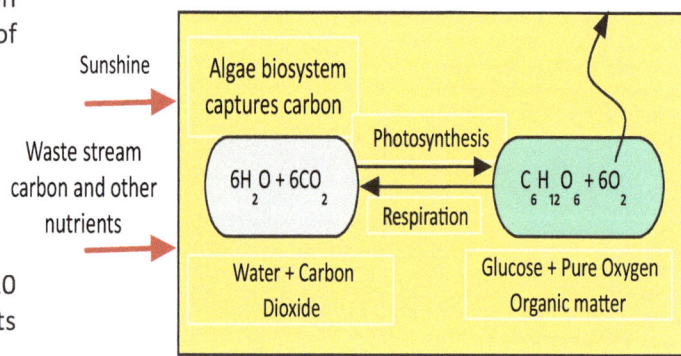

Alternative photon sources provide a critical cultivation bridge in areas that do not have the benefit of many clear sunny days. Vertical farms use LEDs and other non-direct solar sources to deliver photons to the crops.

Single-celled biofactory

Algae provide nature with the fastest growing plant on the planet. Algae do not have to waste energy on roots, leaves and stems.

The single-celled algae biofactory serves as the Ecolanda workhorse. The single algae cell appears quite simple. However, billions of years of evolution have enabled the tiny plant to create a simple but effective array of biomechanisms.

These features act as an integrated biofactory for the production of thousands of bioproducts.

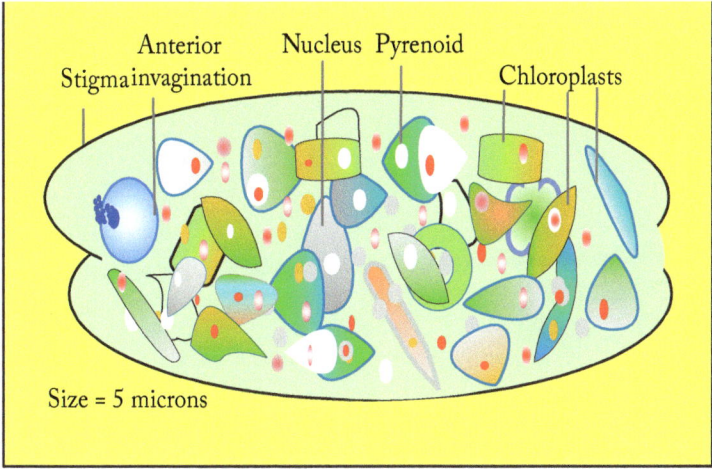

Single-Celled Algae Biofactory

Terrestrial plants uptake nutrients through their roots that flow through their stems and leaves. Rooted plants cannot be grown in wastewater. Wastewater often contains lots of salt ions that would clog their circulatory system in their roots and kill the plant. Algae have no roots so salt ions are not a problem.

The Nrich BioRenew process, right, biocycles waste stream nutrients from polluted air, wastewater, botanical sources, coal mine powder or coal dust. Global companies currently use algae to remove pollutants from municipal, industrial and other waste streams.

Recovering nutrients with land plants would be a loser's game. Terrestrial crops waste over half the applied nutrients, which makes the concept infeasible. Algae use nutrients in the culture very efficiently. The soft green biomass makes it relatively easy for growers to extract compounds.

Dormancy

Terrestrial plants are extremely limited because they are constrained by their roots. Land crops had to adapt and grow even if the soil lacked certain nutrients. This "hidden hunger" effect means that field crops may look big and bright.

However, they are likely to lack nutrients that were unavailable in the soil. When terrestrial plants evolved from algae 500 million years ago, they made many sacrifices and left behind extensive capabilities, including the ability to go dormant. If a field crops lack inputs, the plants do not have a choice between dormancy and death. Field crops lacking water or extreme heat or cold, die and die quickly.

Algae do not exhibit hidden hunger because they can go dormant for decades. Each algae cell assimilates the full set of essential nutrients. When the first nutrient is unavailable in the culture, algae growth stops.

The cells remaining in the culture are loaded with nutrients, but growth and propagation pause. Each cell carries the full nutrient set, which is why algae nutralence significantly benefits consumers whether eaters are people, animals or plants.

Growers may stress an algae culture by withholding nutrients or changing culture parameters. Algae biofactories respond by making new, often valuable compounds such as astaxanthin.

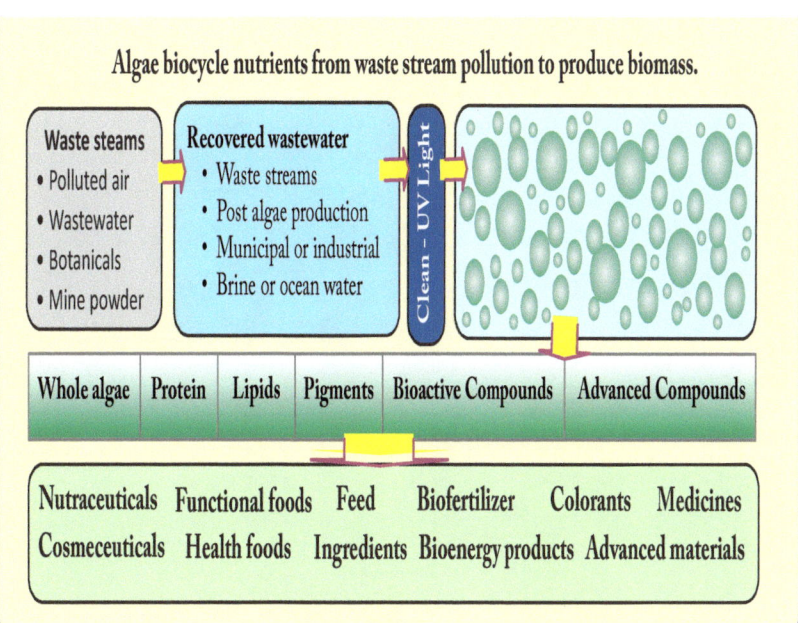

Speed of growth

How do algae grow so fast? Algae distribute the limited energy they receive from the sun very differently than land crops. Field grains are multicellular organisms that invest most their energy in non-food components necessary for survival in terrestrial ecosystems.

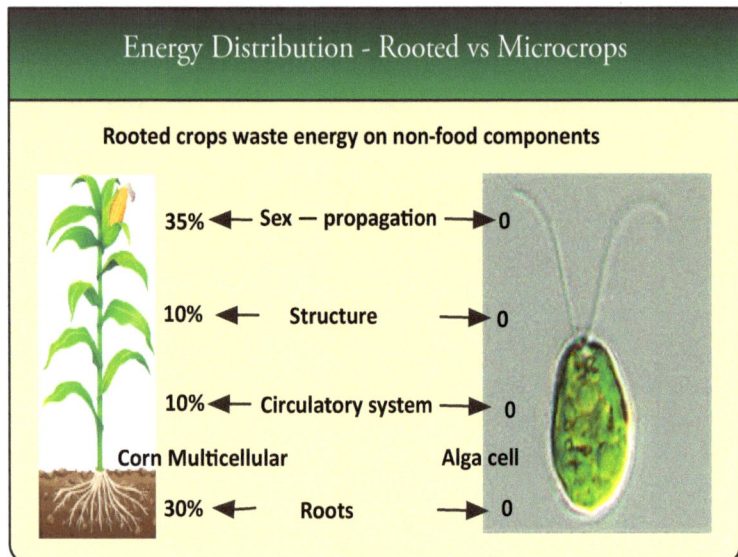

Land plants must anchor themselves with roots and withstand the vulgarities of storms, winds and water. Over 90% of biomass growth – roots, circulation system, structure and even sexual apparatus – consumes huge amounts of energy but grows no food. Consequently, a corn plant

Algae are single-celled, water-based plants with no roots and no circulatory systems. Not only do algae plants grow 50 times faster than field grains, but the biomass also contains over 90% food and other valuable compounds.

The next section examines the water-energy nexus.

11. Water Energy Nexus

Water is energy intensive.
Energy is water intensive

<u>Water is the driving force of all nature.</u>

— <u>*Leonardo da Vinci*</u>

Ecolanda treats water with the respect it deserves, as the soul of the earth. The lifecycle and the water cycle are one. Ecolanda growers and citizens treat water as our most precious natural resource.

SCAD treats water, the most critical natural resource, with a novel three-layer strategy.

1. **Save** – save water with advanced technologies and smart applications that use 80% or less water compared with industrial methods.

2. **Substitute** – save potable blue water by substituting non-potable water for extensive bioproduct cultivation.

3. **Sparkle** – clean waste and other non-potable water, recover and repurpose the nutrients and create sparkling blue water.

These three strategies drive all Ecolanda design decisions because water conservation success requires rigorous systems integration.

Intensive mechanical agriculture often uses 80 to 90% of a community's fresh blue water.[136] Much of the water goes onto cropland to produce animal feed. A single pound of beef uses 1800 gallons of water.[137] Most irrigation water flows onto cropland for animal feed or biofuels.

American farmers use nearly 3 trillion gallons of water annually to grow corn for biofuel production.[138] Most countries do not produce biofuels because lifecycle analysis shows corn ethanol production uses more energy than the fuel delivers.[139] Farmers in the US produce biofuels only because they receive substantial subsidies. Policy makers ignore the massive water loss and ecological costs to society.

Strategies to save water

Water management begins with saving water and eliminating runoff from fields which carries agri-chemicals into public waterways.

Ecolanda Goal: Reduce net water consumption by 80% compared with industrial agriculture and other social and economic development.

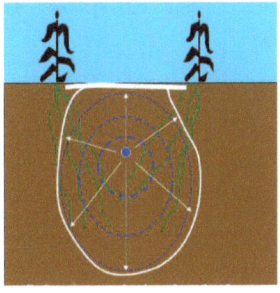

SCAD employs extensive sensors and monitors to manage and control water preservation in agri-energy, waste and the ecocity. Field crops are irrigated by sub-surface drip. These systems save about 80% over surface irrigation. Field systems irrigation typically loses about 50% of the water from evaporation.

Sub-surface drip systems do a better job of spreading moisture throughout the rhizophore, root zone, and avoiding evaporation. Improved moisture dispersion increases root depth and strength, improving crop growth and vitality.

Cropland typically varies significantly in moisture retention according to the type of soil in each micro-geology. SCAD installs remote soil sensors approximately every six hectares to ensure all areas in a field receive precisely the water needed to support healthy crop growth.

Sub-surface drip irrigation avoids runoff due to careful moisture control. In areas where annual monsoons or typhoons are likely to overwhelm fields with driving rain, catchments reservoirs are built to save water for future use.

Ecolanda buildings are designed with catchments that vary from simple rain barrels to large storage tanks. Water-smart architecture employs a variety of retention technologies that save water that would otherwise become runoff.

Ecolanda uses Controlled Environmental Agriculture, CEA, extensively. Hydroponics and aeroponics save 80 to 90% of water compared with surface irrigation.

Vertical farms

SCAD CEA food production in vertical farms save 90% of cropland, 90% of inorganic fertiliser and 95% or more of industrial pesticides and poisons. SCAD CEA allows no emissions or pollution of air, water, soils or ecosystems.

Field crop water loss occurs from about 50/50 evaporation from the soil and plant transpiration.

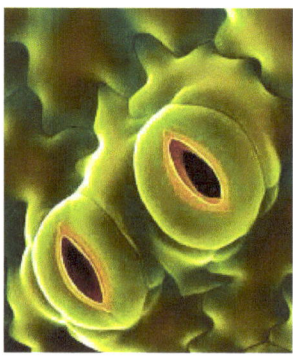

Transpiration keeps plants cool by allowing water droplets to exit through their stomas, left, on leaves and stems. Plants lose 90% of their water from transpiration.[140]

Ecolanda uses vertical farms to catch and reuse both evaporation and transpiration residual moisture. Moisture rises to the ceiling, which may be multiple stories high. Sensors identify when enough moisture has accumulated, which switches on a tiny pump that flows excess moisture to a storage catchment for reuse.

SCAD uses CEA in vertical farms to produce 90% of aquaculture feed, which saves millions of litres of irrigation water that would be lost by field crops. Fish thrive on algae biofeed which delviers superior nutralence and enhances fish growth and development. SCAD cultivates algae aquaculture feed using zero blue water.

Vertical farms are used to cultivate about 60% of the feed for meat and dairy animals. These animals need some ruffage for good digestion, which come primarily from field grains. The majority of animal nutrients are delivered by CEA feed crops such as alfalfa, barley shoots and high-nutralence algae biofeed.

CEA irrigation designed to biocycle water for recovery may reuse the same water multiple times for different crops. Aquaculture water carrying rich fish excrement flows to CEA and provides the nutrients for vegetables and fruit.

Zero-carbon vertical farm

Conventional greenhouses are not airtight and lose considerable energy and water. A biofoam greenhouse offers 90% less water loss, which allows maximum recovery from crop evaporation and transpiration.

SCAD is working on highly insulated vertical farm construction materials such as biofoam that make substantial savings to both energy and water. Vertical farms illustrate space efficiency.

Construction-grade biofoam protects from typhoons, earthquakes, fire and water invasion. This biodegradable material saves 80% on utilities because heating and cooling energy, as well as transpiration water stays within the building.

Ecolanda plans to use biofoam extensively for many types of buildings across the five sectors. SCAD makes the material from algae oils and fibres cultivated in algae biosystems. Similar biosystems create biodegradable bioplastics, biofilms, packaging and construction materials.

How are Ecolanda biomaterials relevant to water? Finding substitutes for water-expensive materials saves enormous amounts of blue water.

Over 70% of people globally live in concrete structures.[141] Concrete consumes enormous amounts of water and emits 5% of total GHG. Concrete is the second most consumed substance globally, after water.

Each cubic meter of concrete uses 160 Kg of water added to the Portland cement, sand and aggregate mixture. Extensive additional water is consumed in mining inputs and curing cement.

SCAD plans to create biodegradable building materials using zero potable water. Zero water, zero carbon building create a very positive eco-footprint.

Blue water substitutes

People, animals and most land crops require fresh water. Algae cultivation may use waste, brine, brackish or ocean water. Blue water substitutes save millions of litres of blue water for other purposes.

SCAD works with several universities that are developing a suite of feed and energy field crops that produce with saline water. These crops can grow in soil too salty for modern field grains.

Blue water saving and substitutes are excellent ways to conserve blue water. Ecolanda goes further and creates fresh, drinkable blue water.

Sparkling blue water production

SCAD applies several technologies to transform waste, brine or brackish water to sparkling blue water. Technology selection depends on geography, latitude and the local hydrosphere.

Ecolanda goal: Create 20% more blue water than used in Ecolanda, including the farm and ecocity

Waste streams made from gases, water and biosolids enter the biosystem. Algae go to work using photosynthesis to power the carbon and nutrient capture. Algae grow and propagate using the waste stream nutrients until the water is sparkling clean. Some water systems require extra filtering to assure clean water.

Growers remove the rich organic biomass using filters, centrifuges or gravity settling. Algae biomass follows the path of petroleum refining and gets converted to multiple bioproducts. A coproduct of algae cultivation, sparkling blue water, benefits the community.

SCAD employs several biotechnologies to ensure pathogens are killed before they enter the BioRenew process. Sensors and monitors create continuous records of water quality. Biological water remediation solutions are preferred because they use solar rather than mechanical energy. Biosystems biocycle the carbon and other nutrients, which are lost with mechanical desalinization systems.

Mechanical desalinization plants cost several hundred million dollars to build and millions a year to operate. Desal fails to capture the carbon or other nutrients.

Desal creates severe problems with dumping excessive residual salt. Many coastal estuaries and oceans that may seem ideal for desal contain too many toxic chemicals.

Desal concentrates the toxins, making residual disposal severely damaging to local ecosystems and sea life.

SCAD's BioRenew process tackles the bulk of water treatment as shown in the diagram.

![SCAD BioRenew Wastewater Bioremediation diagram showing Sunshine or LEDs and Waste streams (CO₂ and CH₄, Polluted air, Wastewater, Farm wastes, Municipal waste) feeding into Algae raceways, then to Clarifier → Reclaimed water, Thickener → Blowdown option, Dewater → Food, feed, fertilizer, and Bioproduct conversion → Oils, biofuels, Bioplastics, Green chemicals, with Recycle water — growing media loop]

The BioRenew model shows sparkling blue water as a coproduct of bioproduct production. Open raceways in the diagram allow water evaporation. The same process may occur in insulated vertical farms that recapture and reuse water multiple times. Highly water efficient closed PBRs, below, may be used in place of raceways.

The process of reclaiming water saves extra water. Growers recycle the growing media, the culture containing residual nutrients. New waste streams flow into the BioRenew process replacing the blue water that is reclaimed.

Outdoor raceways grow biomass only while the sun shines. Vertical farms and CEA using LED, laser or fibre optic light may extend production up to about 16 hours a day. Algae cells are work horses, but they are also living organisms and need a rest daily.

The next section examines Ecolanda's smart ecocity.

12. Ecolanda's Smart Ecocity

Like music and art, love of nature is a common language that transcends political or social boundaries.
— Jimmy Carter

Goal: Ecolanda's smart ecocities inspire awe and joy and deliver economic growth, health and happiness for residents.

Ecolanda smart ecocities are designed from their foundation to function as bioeconomies with a robust, sustainable circular economy. Thoughtful architecture and engineering design with systems integration in mind to use entirely green energy and create zero emissions or pollution.

Ecolanda ecocities operate with no or minimal consumption of fossil resources, capture rather than emit carbon and produce 20% extra blue water. Smart land use assures plenty of green space, recreational parks and botanical gardens.

Smart cities accrue social and economic benefits from positive economic growth and lower costs from using resources efficiently. Automation improves reliability and speed while it saves costs.

Intelligent systems and the internet of things, IoT technologies, automate city resources such as communications, power, transportation and water. Connected smart buildings save resources.

Renewable energy eliminates fossil fuel pollution. Connections improves water and energy reliability, performance and allows effective, data-driven decisions. Smart transportation reduces congestion and pollution, improves safety and saves citizen's time.

Sensors monitor utilities and events. They mitigate risks and reduce damage from extreme events. Sustainable ecosystems with zero emissions create a clean city, which increases the standard of living and happiness. These actions lead to economic growth.

Ecocity model

Ecolanda ecocities model the self-sustaining resilient structure and function of natural ecosystems. Ecocity architecture design, below, allows citizens to live comfortably within environmental means. Ecocities eliminate carbon and other wastes and use exclusively renewable energy and resources.

Ecolanda's ecocity stimulates economic growth, reduces hunger and poverty and improves health and vitality for residents. Each ecocity is unique based on geography, history and culture.

Ecocities strive for efficient land use, green public transportation, efficient use of resources and habitat preservation and restoration.

Happy citizens care about their city and work to make additional improvements. Talent flows into smart cities attracted by the vitality and healthy environment. Happy residents enthusiastically drive the economy.

The ecocity houses Ecolanda associates and their families. Residents include others who desire to live, work and play in the healthy, safe and vibrant community. The ecocity adopts best practices and learning from ecocities currently in designing, planning, building and operations.

SCAD has assembled a group of world-class architects to design each Ecolanda smart ecocity. Architectural designs include inputs from local, regional and national stakeholders. Participation ensures designs reflect the community's vision and values. Architectural design aligns with national and local culture, history and art.

Ecocity benefits

The smart ecocity benefits from the attributes described in the graphic below.

SCAD pursues an 80% goal for local production of food, energy and blue water used in the ecocity. The blue water comes from ecocity wastewater cleaned and biocycled to clean blue water.

The ecocity offers several novel features besides affordable housing and green transportation. SCAD plans an international sports training facility for young aspiring athletes. Several medical and health facilities will draw people who want to improve their diet, exercise and lifestyles. Abundant green space engages residents in recreation and helps restore biodiversity.

SCAD Education and Training

SCAD Ecolanda has a strong commitment to early education and life-long learning.

Ecocity infrastructure supports education, social, family, recreation and religious needs. Best practices guide social, community, environmental and e-governance. The community publishes its shared values to create a sense of belonging for everyone.

Education and training options offer both in-person and distance learning options. Staff provide skills training for thousands of Ecolanda associates.

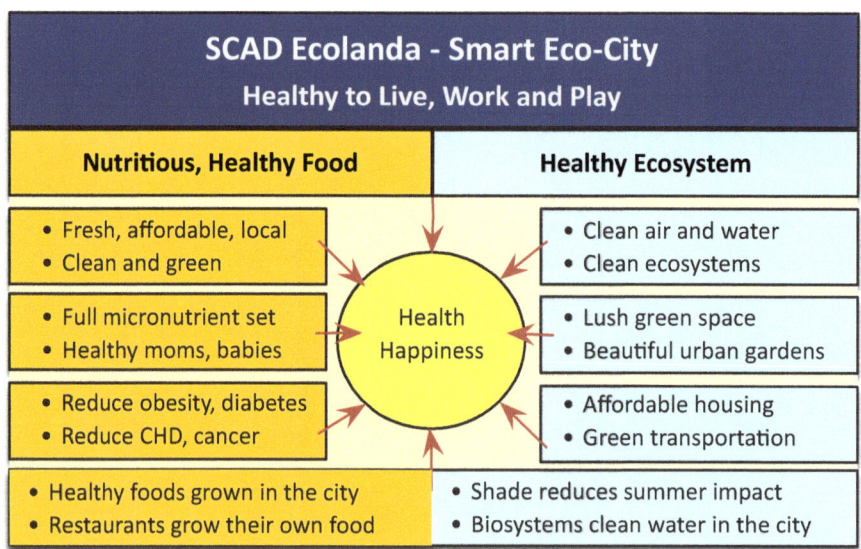

These include agri-energy-water operational and experiential courses. Staff deliver education on basic, intermediate and advanced sustainable earth-friendly technologies as well as sustainable lifestyles.

SCAD plans include building and staffing a pre-school, primary and high school in the ecocity. Staff partner with appropriate stakeholders in building school curriculums, including for colleges and universities. Staff also partner to assist with building a health, safety and sustainable lifestyle curriculum.

Eco-tourism

The Ecolanda agri-energy farm includes an excellent training facility for associates and local people with a desire to learn about sustainable agri-energy. The learning centre will become an eco-tourism attraction for people globally who wish to learn biosolutions and how to build and operate fascinating sustainable technologies.

The learning centre will introduce new crops and better ways of cultivating crops. Emphasis on practical learning ensures that students are able to take away capabilities for their own farms, homes, apartments and urban gardens.

Microcrop cultivation requires considerable attention and knowledge. SCAD will create an online network to assist growers remotely using their cell phone or PC.

Sensors will transmit daily culture data from the grower to a central support centre. A team of SCAD experts, academics and volunteers will answer questions. The team may recommend specific strategies to avoid pests and to successfully grow microcrops. The team will also assist with down-stream processing and developing markets for local bioproducts.

Indigenous agri-techniques offer value. A cross-functional team will learn agri-methods that have been used for centuries from indigenous people. The team will create a curriculum set for local people to learn modern sustainable agricultural techniques they can practice on their farms and communities.

Sustainable agriculture learning centre

Entrepreneurial Park

The Ecolanda community benefits from a diverse entrepreneurial park. The park includes a technology incubator that mentors and launches vibrant companies that take novel Ecolanda technologies to the next level. Focus areas are shown in the graphic.

Ecolanda advanced technologies are used across the entire project and include:

- Internet of things - serves as SCAD's collection, monitoring and data first-line data analysis tool.
- BioRenew - biocycles carbon and other nutrients.
- Abundance agricultural methods - applies BioRenew to preserve natural resources and avoid overconsumption and waste.
- Smart agri - precision agriculture with smart sensors, monitors, data capture and analysis.
- Biofertiliser - biocycles waste nutrients from air, gas, water and biosolids into clean, high nutralence plant food.
- Biofeed - biocycles waste nutrients to recover, recycle and repurpose nutrients into clean, highly nutritious animal feed.
- Bioenergy products - coal dust bonded to algae; briquettes burn hot and clean.
- Bioenergy storage - aluminium ion battery, green hydrogen and fuel cells.
- Emerald H_2 and Emerald Ammonia production - energy products biocycled from waste streams.
- Smart energy storage - local energy storage in aluminium batteries.
- Smart microgrids – green energy produced and stored locally avoids transmission loss.
- Smart eco-city - connected smart city produces its own food, energy and water locally and avoids waste and pollution.

The ecopark creates new jobs and strengthens the local, regional and national economy. SCAD devotes funds and advisors to indigenous people who want to build-out ideas to benefit their communities.

The Ecolanda entrepreneurial park operates as a public—private partnership. Distance education and skills training are available for people at all levels. Student have access to internships, mentors and special projects.

Ecolanda Ecocity Entrepreneurial Park	
Novel batteries	LED lighting
Robotic systems	3D printing
Green energy systems	Smart cities
Smart sensors and big data	Market research
IoT and advanced analytics	Smart transportation
Ecology and econometrics	Sustainable systems
Abundance growing methods	Aquaculture methods
Renewable building materials	New food development
Renewable energy development	Hydroponics production
Zero emissions waste management	Advanced agri-production

Robust internal and external vendor relationships support new ventures with grants-in-kind, new technologies, and experienced coaching. Multi-disciplinary teams elevate new project and assist with the soft and hard launch.

Integrated sensors, monitors, metrics and reporting using dashboards provide insight. These continuous updates and connections support learning and moderate risk. The launch-pad maintains strong ties and engagement with national and global academics.

Smart methods such as simulations and computer modelling are used for prototypes. These actions save considerable time and cost. Guidance for market testing assists potential entrepreneurs such as taste testing for freedom foods, which are the topic of the next section.

13. Freedom Foods

Make a habit of two things: to help; or at least to do no harm. – Hippocrates

Freedom foods help people, animals and plants and do no harm to the environment. Freedom foods are cultivated with abundance methods that preserve non-renewable resources for future generations by avoiding extraction.

Freedom foods are special foods made from algae and other microcrops. They are extraordinary because they are the first food designed to grow free from consumption of fossil resources and free from waste and pollution.

Freedom foods are unique in that they consume zero fertile cropland, fresh blue water, fossil fuels, inorganic mined fertiliser, pesticides or other agri-poisons. Every ton of freedom food captures and repurposes two tons of CO_2.

Freedom foods give consumers a free choice to choose healthier foods for themselves and their families as well as their environment.

Higher nutralence

Freedom foods deliver higher nutralence which enhances health for consumers whether they be humans, animals or plants. Micronutrients are essential to sustain life, cognitive and physiological function. Globally, micronutrient deficiencies, MNDs are pervasive. In America, 85% of citizens of all ages suffer from micronutrient deficiencies.[142]

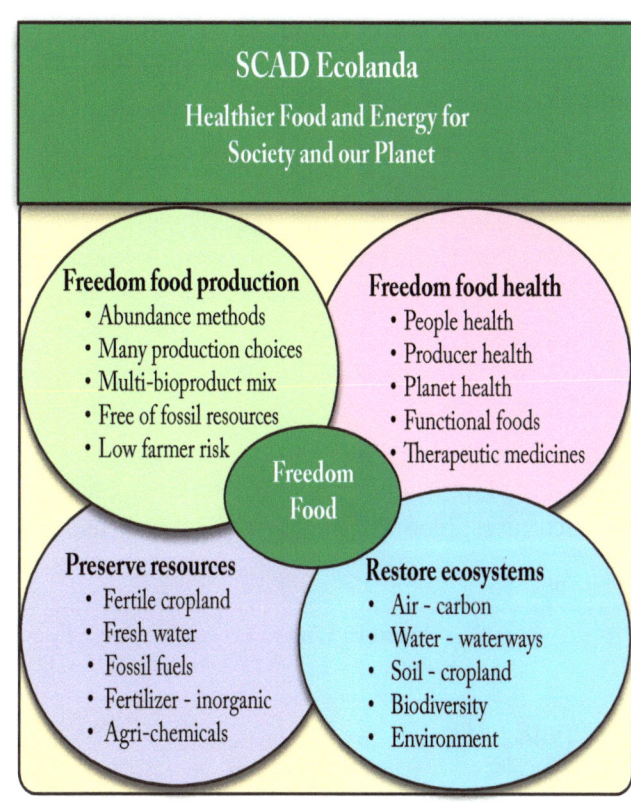

Pregnant women and their children under 5 years are at highest risk. MNDs cause cruelly weak physical development, intellectual, respiratory, vision impairments and perinatal complications. Iron, iodine, folate, vitamin A, and zinc deficiencies are extremely widespread.

Sperm counts have declined 59% over the last 40 years for men in North America, Europe and Australia.[143] MNDs are the most likely culprit.

MNDs put people of all ages at significantly increased risk of disease, premature disability and death. The elderly too often suffer from premature declines in cognitive, neurological, heart, lung and vision faculties due to MNDs.

Freedom foods can end MNDs in a few weeks. Each bite delivers a full set of micronutrients. [Biocycled nutrients](#) provide superior nutrition, micronutrients, bioactive compounds, vitamins, minerals and trace elements.[144] Their high nutralence delivers twice the protein and four times the total nutrients per bite of industrial foods. Freedom foods offer 20 times more natural biodiversity than legacy foods.

Most freedom foods contain dozens of bioactive compounds that are not available in IMA foods. Bioactive compounds protect against disease and fight diseases when they attack.

Freedom foods are composed of microcrops like algae with tiny cells. "Small becomes large" when measuring bioavailability. Bioavailability provides a metric for gut assimilation. These foods supply dozens of healthy food ingredients that are used to fortify and improve legacy foods. Nutrients go into functional foods that function to enhance health and vitality.

Fossil foods depend on fossil natural resources for production. They are often grown in GMO monocultures, (genetically modified organisms.)

A Fossil Food Diet imposes health liabilities.

GMO monocultures means that the plants are essentially clones of one another. They all share the same strengths and weakness.

Monocultures put an entire crop at risk from a single pest vector. Monocultures put farmers at risk as they must use specific herbicides, pesticides and other poisons. GMO crops impose extra costs on farmers as these crops require more cultivation, water, fertiliser and agri-poisons.

The Freedom Foods Diet offers significant health benefits beyond higher nutralence. Freedom foods made from microorganisms do not impose the high fat and cholesterol common in food made from field grains.

Rice, wheat, maize and other grains taste hard, bitter and starchy. Food processors add extensive sugar and salt. Freedom foods are tasty without adding extra sugar and salt.

Plant-based meat and dairy freedom foods avoid the curse of fossil meats, high fat and cholesterol. Freedom foods use no GMO, which avoids the comprehensive concerns about GMO crops and health for growers, consumers and the environment.

Western Diet of Processed Fossil Foods Manufactured GMO Foods with Hidden Hunger		
Processed foods • High fat • High cholesterol • Medium protein • High sugar • High salt	**Food types** • Highly processed • High animal dairy • High animal meat • High GMO • Pesticide residuals	**Chronically low** • Low fiber • Low nutrient quality • Low nutrient density • Low bioavailability • Zero bioactive compounds

Fossil foods carry pesticide contaminants which can damage the brains and major organs of unborn and newborn children. Freedom food growers use no agri-poisons, which zeros-out pesticide residuals on and in food.

Contrast Fossil Foods with Freedom Foods.

Freedom Foods Diet Whole, High-Nutralence Natural Foods		
Whole foods • Low fat • Low cholesterol • High protein • Low sugar • Low salt	**Food types** • Whole foods • Plant-based dairy • Plant-based meat • No GMO • No pesticide residuals	**High nutralence** • High fiber • High quality • High nutrient density • High bioavailability • High bioactive compounds

Many freedom foods are high in fiber, which helps with digestion and gut health. Microcrop meat products may deliver twice the protein per bite as fossil meats. Freedom foods deliver over five times more nutrients per bite and 10 times more nutrient diversity.

Save Biodiversity

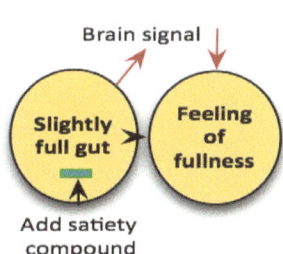

Fossil and Freedom Foods

Freedom foods

Microcrops
- Rootless
- Microorganisms
- Weather independence
- Fossil independence

Minicrops
- Vegetables and herbs
- Fruits and nuts
- Aquaculture, hydroponics
- Low fossil dependence
- Grown in vertical farms

Fossil foods

Fossil foods
- Field crops with roots
- GMO monocultures
- Animal meat & dairy
- Industrial agriculture
- Fossil resource dependent

Therapeutics

Many freedom foods deliver nutraceuticals and therapeutics that treat or avoid disease vectors. These foods deliver superior aroma, colour, texture, and taste. They have less than half the fat and cholesterol than field grain foods.

Abundance growing methods reduce production risk for growers while diminishing waste and costs. Growers have no risky exposure to heavy farm equipment, diesel fumes, dust or agricultural poisons. Rural communities are spared from agri-emissions and pollution.

Microfarms can provide climate independent food production year-round in a city, even after a natural disaster.[145]

Nosh

Nosh refers to eating greedily. Nosh has the connotation of eating on the sly, between meals, or possibly sweets in secret. Eating foods with empty calories leaves the brain with no satiety, the feeling of fullness. The old Cracker Jack box expressed nosh perfectly, "The more you eat, the more you want!"

This was a successful tagline when most people were hungry and thin. Borden's and then Frito-Lay had to change their advertising for modern obese children.

Obese children listen to their nosh signal that tells them to eat more food, more often. If the food contains empty calories, their little brains continue to signal: "More, more, more."

Rimonabant or possibly other drugs can help children, but the brain signal side-effects are far too dangerous, except for morbid obesity.[146] Algae food may provide a safe therapeutic solution.

Algae compounds provide an array of medical benefits for children plagued with nosh signals that lead to obesity and diabetes. Two unique strategies may be called fill-gut and gut-full signaling.

The **fill-gut** strategy adds a few grams of dried algae eaten early in a normal meal, possibly as sprinkles on a salad. The algae expand and fills the stomach.

Alginates can absorb 300 times their weight in water, which quickly fill the gut and suppresses appetite by sending the gut full signal to the brain.

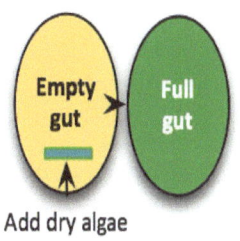

The **gut-full** signals work because algae compounds activate the stomach's natural satiety signals.[147] Satiety signals an immediate feeling of fullness, which tells the eater to stop eating naturally. Algae satiety signals the brain and quashes the nosh feeling.

The nosh feeling hits children especially hard, which is why so many children are overweight.

Many types of algae contain alginates that quash the nosh directive. These foods expand in the gut creating a feeling of fullness that blocks nosh. Brain signals help young people and adults stay at healthy weight.

Algae antivirals use ingenious strategies attack and defeat viruses.

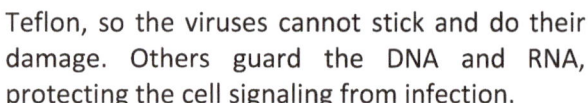

Some compounds make cell walls like Teflon, so the viruses cannot stick and do their damage. Others guard the DNA and RNA, protecting the cell signaling from infection.

Algae compounds block viruses from releasing genetic material or create disruption. Another algae compound strategy inhibits the virus from reproduction. Algae biostimulate the cell to produce macrophages that engulf the virus. The body senses the surrounded virus as a waste material and sluffs it off naturally from the body.

Functional nutrition

Algae biomass produces proteins, lipids, carbohydrates and many specialty compounds. Proteins are large organic compounds made of amino acids arranged in a linear chain connected by peptide bonds. The plant's genetic code determines the sequence of the amino acids, but nutrient limitations may cause changes to the production of amino acids.

Algae's vast biodiversity offer excellent potential for the cultivation of high valuable molecules. Most of these are not easily produced in land plants. Many high- value recombinant proteins are therapeutic proteins. Recombinant proteins may be used to fortify nutritional value.

Recombinant proteins often retain their biological activity when they are ingested, thus improving human health. These functional proteins like the colostrum protein osteopontin, are naturally present in breast milk. They have been shown to impact brain development and immune system function.

Another functional protein found in breast milk, Immunoglobin A, is an antibody found in most body secretions due to is antimicrobial activity.

Most proteins are enzymes that catalyze biochemical reactions and plant metabolism. Other proteins maintain cell shape and provide signaling functions.

Protein plays a crucial role in the human diet, providing most of the nitrogen humans need. Proteins deliver a subset of amino acids that cannot be synthesized by the human body and those need to be supplied in the diet. These essential amino acids are histidine, isoleucine, leucine, lysine, methionine, phenylalanine, threonine, tryptophan and valine.

Other "conditionally essential" amino acids may not be synthesized properly by the body. These include arginine, cysteine, glutamine, glycine, proline and tyrosine. Global hunger is characterized by Protein-Energy Malnutrition (PEM). Deficient intake of essential amino acids reduces total energy and causes a series of extremely dangerous conditions.

PEM can be solved with a new source of inexpensive and balanced protein from algae and other microcrops.

Some algae have a very high percentage of their dry biomass as protein. Species like *Arthrospira platensis, spirulina,* have up to 70% of their biomass as protein content. A short 8-week course of 3g a day of spirulina have been shown to eliminate PEM and other micronutrient deficiencies in children and adults.

Lipids are long carbon chain molecules. Lipids store energy for the plant and serve as the structural components for cell membranes. Lipids are an indispensable component of cells and are precursors of many essential molecules. Lipids are crucial for the human diet.

Some algae accumulate lipids to 70% of the dry biomass. Similar to essential amino acids, some lipids are essential, including a-Linolenic acid and

Linoleic acid. Several algae lipids have proven to have a positive impact on human heart, brain and circulatory systems. Long chained omega-3 fatty acids docosahexaenoic acid (DHA) and eicosapentaenoic acid (EPA) are not naturally synthesized in animals or land plants. They must be acquired through diet.

The traditional source of such nutrients in human diets has been seafood in general. Fish contain omega-3 fatty acids because they consume plankton and algae as part of their diet. The essential long chain polyunsaturated fatty acids are produced in algae.

Many fish stocks have been over-harvested to remove the omega-3s. The daily omega-3 dose for humans may consume 16 sardines or other small oily fish. Loss of small fish stocks creates a cascade of hunger in the ocean which drives down large fish growth and numbers. Consuming algae-based omega-3 fatty acids offer a healthy alternative for humans and fish.

Algae provide an excellent source of vitamin A, vitamin B complex and vitamin E. Other algae nutrients that have a positive impact in human health are antioxidants lycopene, b-carotene, and astaxanthin and polysaccharides, beta-glucans. Beta glucans are a soluble dietary fiber that is strongly linked to improving cholesterol levels and boosting heart health.

Algae pigments are different from land plant and synthetic colourants because they provide nutrition. B-carotene is transformed in the human body into the essential vitamin-A. Astaxanthin, also a carotenoid, has a distinctive red colour used in animal feeds to confer a deep yellow colour in chicken egg yolks and the red colour used in farmed-raised salmon.

Farm-raised salmon fed field grains are not marketable because they lack the pink salmon colour.

Many algae species have therapeutic properties that improve health and prevent or treat disease. Algae bioactive compounds have proven benefits against degenerative metabolic disorders.

Metabolic syndrome, which is highly correlated with the Western Diet of fossil-based foods, imposes severe health conditions throughout life. The metabolic syndrome, which is very common globally, can be avoided with a diet that includes microcrops.

Bioactive compounds

Freedom food nutrition offers over 300 bioactive compounds that provide organisms with various types of shields against disease vectors. Industrial agriculture does not deliver bioactive compounds in produce. Bioactive compounds developed in algae-based foods over their 3.5 billion years of evolution. Land plants evolved from algae about 500 million years ago.

Primary bioactive compounds in algae
- Antioxidants
- Soluble dietary fibers
- Proteins and amino acids
- Minerals and trace elements
- Vitamins and phytochemicals
- Polyunsaturated fatty acids

Bioctive compounds for degenerative metabolic disorders:
- Sulphated polysaccharides
- Phlorotannins
- Vitamins and minerals
- Carotenoids (e.g. fucoxanthin)
- Peptides and sulfolipids

Improve health, prevent or treat disease
- Anticancer
- Antioxidant
- Antiobesity
- Antidiabetic
- Antihypertensive
- Antihyperlipidemic,
- Anticoagulant
- Anti-inflammatory

Other bioactive compounds:
- Immunomodulatory
- Antiestrogenic
- Thyroid stimulating
- Neuroprotective
- Antiviral
- Antifungal
- Antibacterial
- Tissue healing

Sea vegetables, macroalgae — Green algae — Blue-green algae, cyanobacteria

The move to roots, stems and terrestrial stressors forced land plants to leave behind any "excess" baggage. Most bioactive compounds either do not exist or are in such sparse amounts they are not measurable in land plants.

A land plant such as maize evolves very slowly because it produces only one crop a year. This means any mutations or even hybridization trials require a decade or more.

Microcrops like algae create a new crop of offspring daily, every day the plant has sufficient sunshine, moisture and nutrients.

Algae, including macro (seaweed), micro (green) and cyanobacteria (blue green), had to survive in extremely rigorous early-earth conditions. Algae form the foundation of the food chain which means that every consumer higher on the food chain viewed algae as food. Algae created a brilliant strategy to contend with all these ravenous consumers – reproduce faster than consumers can eat.

Algae survived every threat vector known as human, animal and plant diseases. These wise plants developed bioactive compounds to neutralize disease threats. The beauty of algae-based food, feed and biofertiliser is that when consumers assimilate the superior nutralence, they receive bioactive compound protection.

Nrich infuses functional foods naturally with an extraordinary set of macro and micronutrients, plus vitamins, minerals and trace elements that are not available in industrial foods. Nrich functional foods provide natural solutions that maximize health and vitality.

Industrial production of fibers and pigments creates massive pollution two air, water and soils. Freedom foods produce strong fibers and more vibrant pigments with zero emissions and zero pollution.

Multiproduct production

Industrial farmers produce a single monocrop commodity. The staple crop makes farm revenue dependent on the commodity price, over which famers have no control.

A maize farmer takes extreme risk as all the crop inputs must be bought and applied to the crop before the maize can be sold. The farmer's risk includes a 10 to 20% chance the crop will fail, and the investment lost. Global warming increases the risk of total or partial crop failure. The maize and sells at the single commodity price, which may be higher than the farmer's inputs.

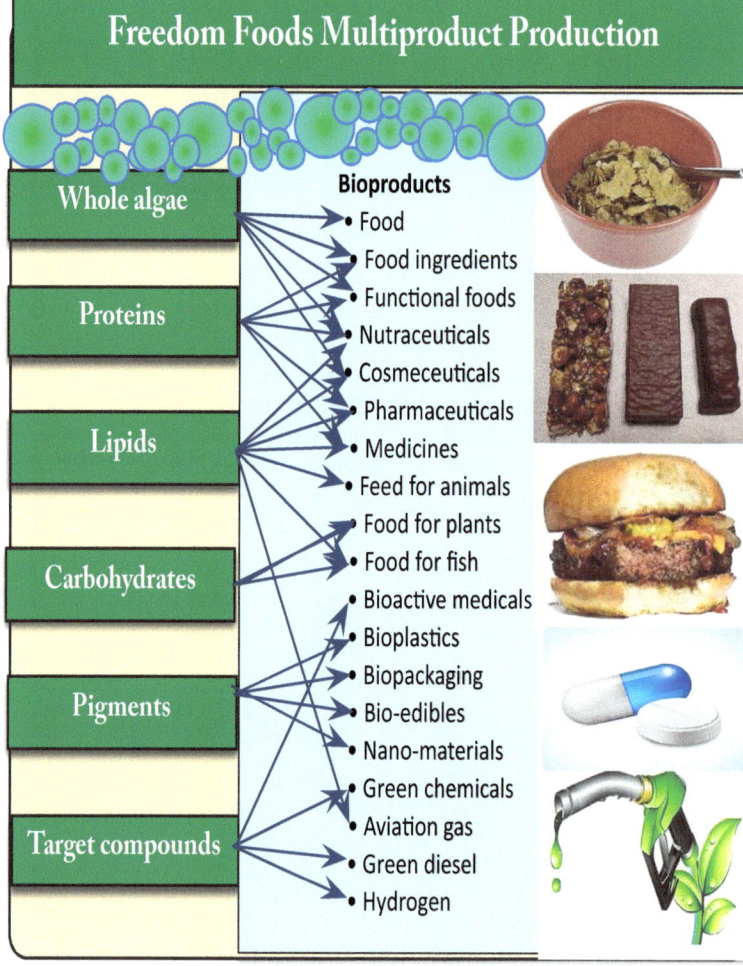

Freedom foods growers do not have the high outlay to pay for crop inputs. They do not have to buy the expensive fossil resources. Their input costs are about 70% less than a field grain farmer.

14. Emerald Flow – Climate Care

No matter how rich you are, you can't get healthy air. – **Ma Jun**

Industrial emissions spread carbon, black soot particulates and other toxins across the entire community. SCAD's Ecolanda commitment:
- Zero emissions from Ecolanda
- Clean GHG and other contaminants from industrial farms equal to 3 times the Ecolanda campus area

Ecolanda designs solutions to reverse climate change. After water, Ecolanda growers and citizens treat our shared atmosphere as our most precious natural resource.

SCAD's emissions management involves a four-layer strategy.

1. **Emerald flow** – reverses carbon flows in our food supply chain from carbon discharge to carbon capture. Abundance methods allow carbon to flow into bioproducts rather than out from industrial crops into the air. This may sound too simple, but the direction of carbon flow may be nature's miracle that can reverse global warming.
2. **Capture emissions** – save the atmosphere with zero carbon emissions from the Ecolanda campus.
3. **Clean air** – clean air with renewable energy, biosystems and direct air capture.
4. **Smart energy products** – invent novel products that replace coal with no black soot particulate pollution.

Each strategy contributes to the promise of clean, fresh air. Emerald Flow differentiates Ecolanda from other development models. The carbon flow from fossil foods flows from food production into our atmosphere. Abundance methods reverse the carbon flow from the air into foods and other bioproducts.

Fossil food flow

How did nature create plentiful food and energy for over 2 billion years consistently and not pollute the atmosphere?

Nature has done her job extremely well. Humans have not. People began fouling our atmosphere 180 years ago with the start of the industrial revolution. Farmers accelerated GHG production in 1950 with the "Green Revolution." Both revolutions were black. They added billions of tons of black carbon to the air we share.

Industrial production applies physical power to force nature to do the producer's will. Physical energy requires mechanical means and all the fossil fuels needed to power heavy equipment. Fossil food production leads to huge carbon waste flows. Fossil foods create a toxic pollution plume that poisons our atmosphere, waterways and rural communities. The enormous fossil food pollution plume is composed of natural resource waste.

The costly discards and massive pollution are fatal flaw in IMA production. Every year industrial pollution to air, water and our shared environment increases, degrading human and ecological health.

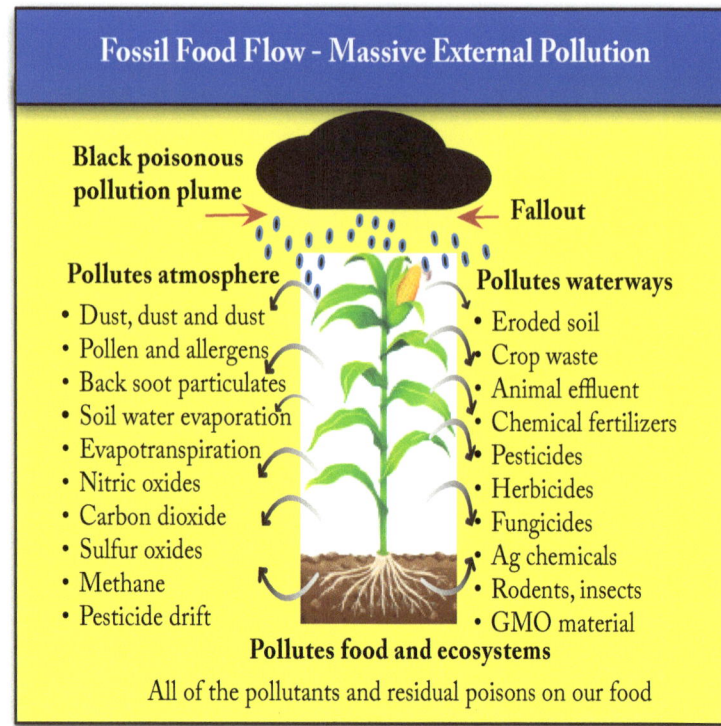

Ecolanda plans to grow a million metric tons of algae annually. This biomass will capture and reuse two million metric tons of carbon. As Ecolanda microcrop production increases, more carbon will flow into bioproducts.

The Emerald flow recycles and repurposes waste carbon and nutrients in microfarm biosystems. Microcrops produce the compounds to make thousands of bioproducts. Nearly any industrial product made today can be made cleaner and faster with Emerald flow in biosystems.

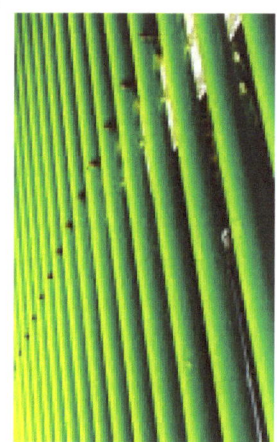

The Emerald Flow works in all types of microfarms, outdoor raceways or in indoor PBRs, right. Healthy algae grow and renew their biomass daily and display the magnificent colour of emeralds.

Emerald Flow

Biological systems reverse the direction of the carbon flow from out to in. The Emerald flow transfers CO_2 in the air to fresh bioproducts.

Abundance agri-energy methods can use the substantial waste created by industrial production and biocycle those wastes into clean food and other useful bioproducts.

Emerald's meaning

Many great civilizations have endowed emeralds with robust meaning, valuing emeralds for nature's power for rebirth and renewal.

Indian mythology, Sanskrit, the beautiful green of Spring and growing things.

Ancient China, Feng Shui, the energy of growth, new beginnings.

Persia, "smaragdus," the earth's rhythm of rebirth and renewal.

Ancient Egypt, the energy of nature and of eternal life.

China, living a good life, happily in harmony with nature.

Aztec and Incas, wealth, prosperity, rebirth and renewal.

Islam, verdant, inspires happiness and assures love from Allah.

Middle Ages clergy, nature's renewal and symbol of divine faith.

The Emerald flow aligns with the emerald's historical meaning of renewal.

Global adoption of abundance agri-energy methods would allow the Emerald flow to reverse climate change in one generation.

Carbon capture

Ecolanda's nature-inspired architecture uses link and sync carbon capture and utilization, CCU, systems. Purposeful design allows each potential carbon waste source to be linked to a microcrop biosystem. Biosystems use the Emerald flow in microfarms to transform the potential pollutant to useful green biomass.

> *Photosynthesis is simultaneously the cheapest and most efficient known solution of all carbon and nutrient cycling technologies.*

SCAD's commitment to CCU includes capturing gasses, liquids and biosolids from animal barns.

Clean air

Ecolanda's fresh air systems capture, recycle and reuse airborne carbon. Biosystems remove and replace the dirty carbon, soot particulates and other toxic GHG with pure oxygen.

Clean air technologies make a first pass at cleaning the air by avoiding emissions and linking smokestacks and obvious carbon sources with biosystems. The air cleaning second wave uses microfarms that take their carbon from the air or stored CO_2. Biosystems often use the bubbles from injected air and CO_2 to mix the culture.

SCAD also uses direct air capture, DAC as a carbon source. This mechanical-biosystem hybrid pulls large amounts of CO_2 directly from the air.

Ecolanda's extensive waste-to-energy systems are strong carbon suppliers to biosystem. The pyrolysis heat and pressure elements create flue gasses that are linked with biosystems for CCU. Several solid waste technologies use microorganisms to generate syngas or green chemicals. Components of both outputs can feed carbon into microfarm cultivation systems.

Mitigate nearby emissions

SCAD does not operate any coal or other fossil fuel power plants. Power plants and their nearby coal mines are terrible emitters of GHG and numerous toxic elements including deadly heavy metals. SCAD makes commitments to capture and recycle stack gasses and create remediation processes for areas destroyed by mines.

Microfarms can mitigate some emissions. Power plants run 24/7 while algae only grow and capture pollutants during daylight. Plants typically emit gasses faster than most biosystems can capture them. Therefore, biosystems offer only a partial solution to power plant emissions.

Many plants burn lignite, the lowest grade of coal. Lignite burns with low energy because the coal is poorly formed and mixed with considerable dirt.

A smart energy bioproduct

Lignite and low-grade coal plants create three problems compared with higher-grade coal. They produce less heat and power even though they burn more material because the coal is laced with contaminants. They emit more GHG and carbon particulates from burning more material. In addition, the lignite is more expensive to mine.

SCAD tackles these challenges in reverse order. Higher mining cost occurs because lignite suffers a 30% loss in mining as the material turns to dust.

These coal fines are not combustible and foul combustion chambers if not separated. The fines create a very expensive waste stream. Coal fines are easily carried on the wind and must be transported and buried before these minute coal particles cause deadly dust storms. SCAD uses a coalgae solution developed at the Nelson Mandela Institute in South Africa.[148] Coal fines are fed into biosystems where algae coat the tiny particles.

When dried and pelletized, the coalgae burn hotter than high-grade coal.

The natural algae coating adds oxygen to the furnace, creating more heat and energy. Higher heat burns cleanly and practically eliminates coal soot particulates. The coalgae trash-to-cash clean technology saves costs, decarbonizes the atmosphere and stops black soot pollution.

Decarbonize everything

SCAD Ecolanda plans to decarbonize production across the board. The goal is not zero carbon, but carbon negative production. SCAD leadership believes it is not enough to prove carbon negative technologies. The value proposition and methods need to be clearly conveyed to world leaders.

SCAD targets the five energy sectors that emit the most carbon: energy production, food supply chain, transportation, buildings and industry. SCAD expands environmental preservation beyond the carbon footprint.

Positive Ecological Footprint

A carbon footprint reports only the management or non-management of carbon emissions. The ecological footprint expands focus to forest, cropland and pasture preservation. It includes the ecological impact of the built environment, land with man-made structures. The eco-footprint also reflects concern about the health of fisheries.

Ecolanda designs production, construction, food, transportation, waste and energy systems to be carbon-neutral or carbon negative. Each sector benefits from smart technologies that create a Positive Ecological Footprint, PEF. Ecolanda agri-energy production comes from renewable sources. All-electric transportation moves on roads, rails and pathways covered with materials that create a PEF.

Building construction uses carbon-neutral and primarily biodegradable materials. Construction materials insulate buildings so effectively they reduce lifetime energy costs by 80%. At the end-of-life, which may be 50 years, the construction materials go through the BioRenew technology where their elements are repurposed into new construction materials or other bioproducts.

The smart ecocity uses exclusively PEF materials, energy and services. The built environment benefits from links that eliminate emissions.

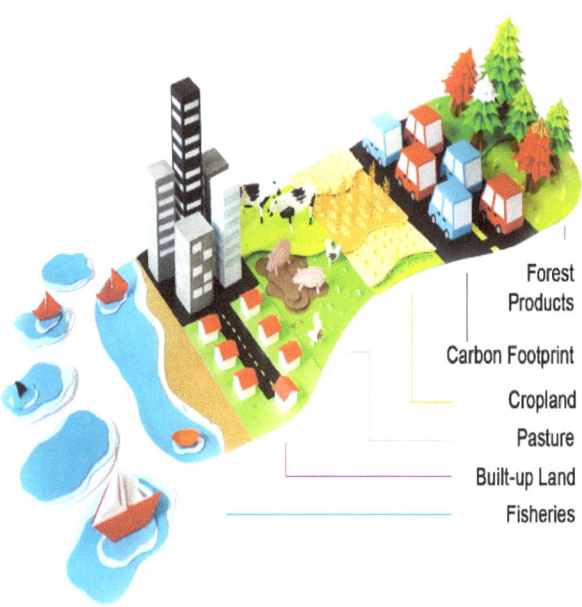

Climate care summary

SCAD's climate strategies eliminate emissions on the Ecolanda campus. The Emerald flow reverses carbon flows from carbon discharge into the atmosphere to carbon capture in biosystems.

Ecolanda agri-energy-waste systems are designed with carbon capture and reuse in mind. SCAD decarbonization includes the following activities.

Carbon sources

The decarbonization graphic summarizes the many carbon sources. CCU requires extensive planning, piping and recovery biosystems. Systems integration pervades all Ecolanda CCU biotechnologies. Smart sensors monitor and report inputs and outputs for carbon and other nutrient recovery, recycling and reuse.

Save Biodiversity

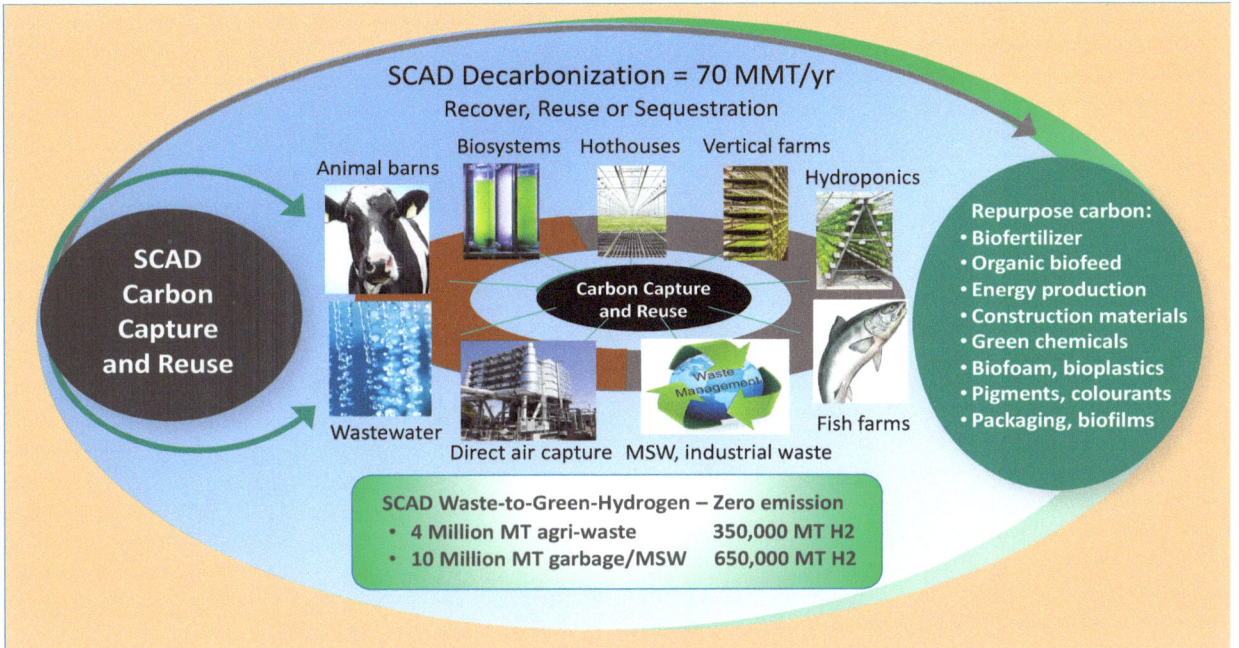

A credible third-party analysis and validates ecological metrics. Industrial Economics, IEc in Cambridge Massachusetts receives live metric feeds. The IEc team analyses the big data sets to determine how much carbon gets captured and how reuse saves ecological damage.

SCAD conscientiously practices ecological transparency. Ecosystem reports go to the host country government such as the Ministry of Interior or Environment and one or more academic institutions.

Ecolanda makes commitments beyond climate to care for all ecosystems. What could be stronger than avoiding carbon release and CCU?

Physical footprint

Each Ecolanda campus varies in size based on land availability and national priorities. Several countries are planning a single campus of over 50,000 hectares. The total land footprint is close but not contiguous.

One country has plans for three sites with the hectares dedicated to each element shown in the table. The agri-energy facility in each is 3,000 hectares. The Nrich facility for microcrops consumes 500 hectares of non-crop land. The SEPS battery plant will fabricate and produce aluminium batteries.

SCAD plans to create a local smart agricultural college in a joint public-private partnership. The Education Centre and Innovation Park are also public-private partnerships.

SCAD Operations Size for Three States

State 100 Ha	1. 8K	2. 11K	3. 12K
Agri-Energy	3,000	3,000	3,000
Nrich Facility	500	500	500
Waste Treatment	1,000	1,000	1,000
Green Hydrogen	250	250	250
Green Ammonia	250	200	250
Smart Ecocity	TBD	TBD	950
Smart Community	500	450	600
Education Centre	50	50	50
Innovation Park	150	500	1,000
SEPS Battery Plant	N/A	N/A	1,000
Agricultural College	100	50	N/A
Energy Crops	2,200	5,000	4,000

Create a clean BioHub

Most ports are noisy, dirty and polluted from heavy use by marine vessels of all types. The heavy mechanical machinery creates constant noise overlayed by traffic and fast-moving cranes.

Cargo and container ships are notorious for sending plumes of carbon into the atmosphere. Their engines are optimized for horsepower, not for clean air efficiency. Many ships leak diesel oil and human contaminants into the harbour waters.

The port types graphic maps waste, air and water pollution consistent with most industrial ports. Several Green Ports are under development with hopes of becoming carbon neutral in decades.

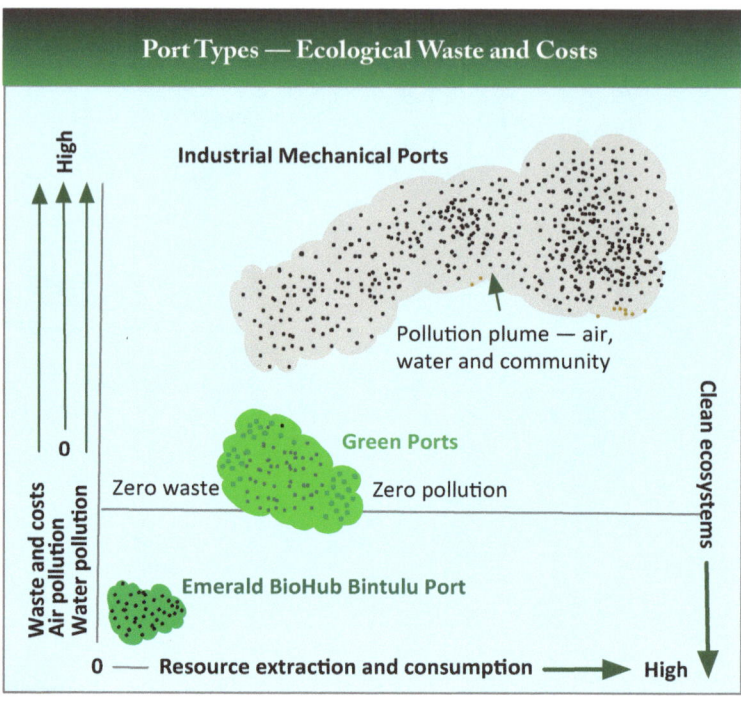

SCAD will build an Emerald BioHub port at Bintulu, Malaysia. The Emerald BioHub port will benefit from high efficiencies and a positive eco-footprint. The port will minimize natural resource extraction, which will occur primarily for constructing green energy systems. The BioHub will create zero waste and zero pollution.

The BioHub will be quiet, using mostly silent electric vehicles and engines. DAC and biosystems on site will captures and reuse millions of metric tons of CO_2.

Biosystems will clean and restore polluted water and local ecosystems. Harbour and coastal ecosystems sequester large amounts of carbon. A kelp forest can sequester up to 20 times more carbon per hectare than land forests. Marine plants that contribute to carbon sequestration. Mangroves, kelps and seagrass live in rich soil. When these plants die, the leaves, branches, roots, and stems get buried underwater in the muck. The low underwater oxygen allows the plant material to stay buried for decades or longer. A kelp forest attract biodiversity that include filter feeders that also cleans the water.

Save Biodiversity

Ecological Restoration		
Air	**Water**	**Soil**
CCU - two tons of CO_2-e for every ton of algae	Produce 20% more fresh water than is used	Bring dead, abandoned soil back to life
Smoke stacks - coal, cement, industry	Clear pollutants from water	Restore the full set of micronutrients
Black soot particulates	Remove pharmaceuticals	Restore soil porosity
Biocycle animal carbon	Remove heavy metals	Recover from salt invasion
Smog - nitric, sulfur and toxic x-oxides	Capture and clean agri-overspill	Restore eroded and worn out soil
Direct air capture - CO_2	Eliminate water pollution	Restore soil biodiversity

15. Emerald Ecological Restoration

Land degradation, biodiversity loss and climate change are three different faces of the same central challenge: the increasingly dangerous impact of our choices on the health of our natural environment. We cannot afford to tackle any one of these three threats in isolation – they each deserve the highest policy priority and must be addressed together.

– Sir Robert Watson, Chair of IPBES[149]

Nrich enriches life by providing superior nutrition for people, animals, plants and ecosystems. SCAD Ecolanda goes further with a commitment to use their novel biological tools to restore degraded and destroyed ecosystems.

SCAD Ecolanda addresses biological ecosystem restoration with a six-layer, 3S3R strategy.

1. **Sanctuary** – create an Ecolanda Sanctuary that practices SCAD's six 3S3R strategies.
2. **Save** – ecosystems from exploitation and extraction of agricultural inputs.
3. **Substitute** – save ecosystems by substituting cultivation methods such as freedom foods and bioproducts that do not exploit cropland or verdant ecosystems.
4. **Recover** – emissions and pollution that would otherwise foul ecosystems.
5. **Rebuild** – fertility and soil structure in degraded ecosystems including macro and microorganisms.
6. **Restore** – flora and fauna to rebuild natural ecosystem biodiversity.

These six strategies work in harmony and drive Ecolanda design decisions. Biological ecosystem restoration does not happen by chance. Biorestoration requires advanced planning, conscientious design and rigorous systems integration.

Ecolanda Sanctuary

SCAD will create a National Sanctuary dedicated to restoring biodiversity. The sanctuary, a public-private partnership, will benefit from, demonstrate and teach SCAD's 3S3R eco-restoration methods.

The Sanctuary will be located near or in the smart eco-city. Visitors will be able to walk trails on 10% of the Sanctuary. The majority will be closed to allow flora and fauna to flourish. Trails will allow robotic rovers to silently tour the Sanctuary. Visitors may use virtual reality headsets to see 360 degrees as if they were riding in the rover.

The Sanctuary will practice "rewilding." This strategy focuses on repairing biodiversity and ecosystem health by restoring natural processes. Rewilding lets nature take over and re-establish balance. The National Sanctuary will be named for a National ecological champion. The sanctuary will include an eco-learning centre where short courses will be available for people of all ages. The learning centre will include a biobank to preserve indigenous seeds and microorganisms.

A National Eco-Restoration Board of Directors will guide and lead the eco-restoration process. The Board will orchestrate return of precious flora and fauna threatened with extinction.

Save

Ecolanda's ecosmart agri-energy production and freedom food saves ecosystems from the need for extraction of industrial agriculture inputs.

SCAD's Nrich biotechnology restores the atmosphere through biocycling carbon and other GHG which mitigates global warming. SCAD engineers off-takes from carbon emitters where the carbon flows to algae biosystems for capture and reuse. SCAD applies direct air capture of CO_2 in addition to Nrich biotechnologies.

Water serves the soul of ecosystems across the environment. Ecolanda operations save millions of liters of blue water, which benefits all stakeholders, including biodiversity.

SCAD's eco-friendly agri-energy methods save rural communities and ecosystems from the nasty plume of GHG and toxic agri-chemicals that invade their air, water and ecosystems.

Substitute

The first step in industrial farming in the Spring is cultivation. Fields are cleared of nearly all life. The term "field cultivation" covers a multitude of sins. The brute force from huge machines rip deep into the soil, killing a majority of the crop-beneficial microorganisms. Cultivation intentionally leaves nothing living to compete with the crop.

Herbicides are applied to kill all competing seeds that may sprout after cultivation. Herbicides kill both seeds and microbial life. Chemical fertilisers then exterminate most the remaining beneficial soil microbes.

Cultivation occurs in the Spring and leaves fields highly vulnerable to erosion from annual winds and rain. Rain and irrigation leave soil wet. Heavy tractors, trucks and wagons repeatedly drive across the field compacting the soil. Years of repeated cropping leave soils devoid of beneficial microbes and severely compacted. Rain does not percolate down through compacted soil.

SCAD prefers no-till farming for broadacre crops to save beneficial microbes and to avoid soil compaction.

Growers use no or minimal herbicides. Nrich biofertilisers provide strong substitutes.

Freedom food cultivation in CEA saves ecosystems by simply not using cropland to produce food and other consumer products. Growing microcrops in vertical farms saves thousands of hectares from the insults inflicted by intensive mechanical agriculture.

SCAD avoids the use of inorganic fertilisers because they kill the symbiotic microorganisms in the soil and too often erode into local waterways.

Ironically, chemical fertilisers must be broken down by the microorganisms in the soil before the agri-chemicals can be assimilated by the crop.

Algae are one of the primary microorganisms that do the breakdown work in soils. Nrich skips the use of mined chemical fertiliser. Algae cells are pre-loaded with the full set of essential nutrients and are delivered to each plant in a form that is immediately bioavailable.

Substituting Nrich biofertiliser for inorganic chemicals saves costs and extensive ecological damage.

Water pollution from fertiliser overspill has created over 410 dead zones in coastal waters and lakes worldwide, affecting an area of 250,000 square kilometers, (95,000 miles^2. This is about the size of New Zealand. Nrich biofertiliser has no or minimal overspill, which saves waterways from eutrophication, loss of oxygen.

Herbicides, fungicides and insecticides used in agriculture cause pregnant mothers to abort or birth newborns with severe developmental disabilities. These toxic compounds cause both acute and often life-threatening poisoning and long-term chronic illness for consumers, farmers, farm animals and rural communities. Pesticide residuals cling to produce, which amplifies health risks for unsuspecting consumers.

Ecolanda uses Nrich biofertilisers that include natural biostimulants and biopesticides. Substituting natural biosolutions for agri-poisons saves millions of families from the tragedy of children that suffer developmental disabilities.

Recover

Industrial methods have polluted the air with CO_2, black soot particulates, dust and toxic agri-chemicals. Nrich biosystems produce healthier food and bioproducts with zero emissions.

Ecolanda uses carbon negative biotechnologies that recover and repurpose rather than emitting carbon and other GHG.

Nrich cleans the hydrosphere through biocycling waste in polluted water and repurposing those elements into fresh, clean bioproducts. Algae biosystems can clean nearly any water source – domestic, industrial, medical and even mining.

Mines and deep wells may use advanced Nrich biotechnology to capture and remove poisonous heavy metals before allowing the biosystem to do its biocycle work. Biosystems recover and recycle wastewater nutrients and toxic compounds that would otherwise pollute ecosystems.

Rebuild

Industrial agriculture makes a classical mistake by intensively farming cropland for decades. Healthy soil contains trillions of living microorganisms. Continuous cultivation and compression from heavy equipment paired with nutrient extraction by crops leaves the soil depleted, compacted and exhausted.

Farmers typically replace only the top three NPK fertilisers – nitrogen, phosphorus and potassium. The other 21 essential macro and micronutrients are ignored. Each crop consumes about half the

NPK nutrients, which are lost to the field with the harvest. Crops continually deplete micronutrients and the humus that holds soil moisture. Nearly all of the NPK residual erodes on wind and water or drop below the root zone where they are no benefit for crops.

Next season, the farmer must buy and apply the NPK fertiliser again with a new round of cultivation. Currently, industrial agriculture has no practical means to rebuild either the micronutrients or the humus.

Nrich biofertiliser rebuilds the soil fertility and structure from the foundation upwards. The subsurface drip system delivers the full set of vital macro and micro-nutrients. Crops respond with higher yields and higher quality produce.

Industrial agriculture systemically compresses soil until it will no longer support crop roots. Nrich can improve soil porosity. Nrich biofertiliser improves soil structure, which allows for stronger and longer root development. Many of the high-nutralence algae cells enter the plants and improve the crop. Indigenous local algae cells continue to grow in the soil creating rich organic matter, humus and plant sugars.

Plant sugars act like honey and attract trillions of additional microorganisms, which create natural symbiotic communities in the looser soil.

Abundance growers improve their fields every year with Nrich biofertiliser. Nrich has the unique capacity to restore soil fertility and even bring dead soil back to life. Growers may address each of the soil maladies shown.

Nrich reconstructs soil quality and structure by increasing nutrient levels and humus in one or two years. Nrich flips compaction to looseness.

Nrich biofertiliser delivers 74 nutrients to the soil. Indigenous algae that delivered the nutrients in algae cells continue to grow in the soil, creating rich organic biomass, humus.

Algae biomass attracts many other microorganisms. The highly diverse micro-community works symbiotically to improve soil structure that benefits fertility and crop growth.

Nrich biofertiliser allows farmers to leave every field better than they found it after every harvest. Algae-based nutrients and humus continue to build layer upon layer of fertile soil with every crop.

Salt invasion

Hundreds of civilizations of all sizes have crashed from a salt invasion.[150] Irrigation water flows long distances and picks up salt. Dissolved salts crystalize in irrigated cropland. When too much salt remains on or in the soil, crops die, children die and then the civilization crashes.

The only historical remedy for soil salt invasion has been to pray for a farmers' rain – long and slow – to flush out the salt. Unfortunately, global climate chaos creates more severe storms that flood rather than soak out salt.

Field trails have demonstrated that Nrich improves porosity in compacted and salty cropland 500%. This metric turns out to be sufficient for the combination of irrigation and rain to flush salt below the root zone in a single season, where it does no crop damage.

Limitation

Nrich can rebuild and save severely eroded soil if the next level below the eroded topsoil is dirt and not rock. Subsurface drip systems can deliver algae biofertiliser and restore nutrients and humus. Fertility restoration may take several years but it can bring dead soil back to life.

Soils eroded down to bedrock are **not fixable** by Nrich methods. This includes surface or mountaintop mines that have scraped off the topsoil.

SCAD offers a solution nature uses for the world's hottest and driest deserts. Nature lays down a thin layer called an algae crust that holds the topsoil layer firmly and resists erosion.[151]

Restore

SCAD Ecolanda restores health to ecosystems and biodiversity. Nrich restores health to degraded ecosystems, which allows the recovery of biodiversity. Healthier nutrition and thriving ecosystems attract extensive biodiversity.

Nature may require a century to create a centimeter of topsoil. Nrich methods may restore fertility in one or two growing seasons. Once fertility is restored, abundance methods ensure continuous soil improvements year over year.

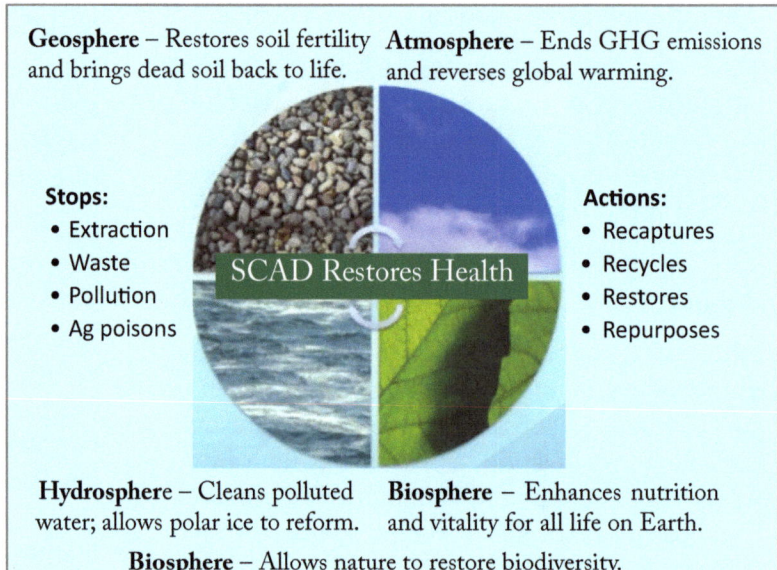

16. Emerald Renaissance

The Emerald Renaissance provides a clean, renewable alternative to fossil resource-based industrial mechanical production.

The industrial revolution produced cheap food and consumer goods. The extremely high ecological cost came from extracting and consuming massive amounts of natural resources and then discarding wastes into the environment. Industrial production imposes an exceptionally high cost on social and economic development and destroys the environment.

The Emerald Renaissance counters the industrial mechanical production with biological cultivation. Biosystems avoid natural resource extraction and consumption and clean rather than foul the environment.

Ecolanda growers are able to produce a wealth of bioproducts using biosystems instead of mechanical production. Growers can cultivate microcrops to produce nearly anything made by industrial means.

Why?

The first question many people ask is:

> If biological systems are superior to industrial mechanical production, why do producers continue to use industrial methods?

Industrial producers find industrial products are cheaper to create than bioproducts. The industrial cost advantage occurs as an artifact of current social policy. Society places no tax on extraction. Industrial producers are able mine resources and the dump or burn their wastes indiscriminately with no tax or cost.

Ironically, instead of paying the full social cost for industrial production, society subsidizes inputs that reduce production costs but increase waste. Industrial producers pay only a small fraction of the social cost for their energy, fuel and water. They pay zero for their waste streams.

Future generations will pay those taxes in the form of much higher costs for scarce resources. Our children and their children will pay extraordinary costs for the impacts of human-caused climate change and polluted air, water and ecosystems.

If society charged industrial producers for fossil resource loss and environmental damage, microcrop bioproducts would be 30% less expensive than industrial products today.

Ecolanda bioproducts

Ecolanda will demonstrate to the world the benefits of biosolutions. Growers will cultivate microcrops for food, feed, fertiliser, green chemicals and energy products. Cultivation will include health products, pharmaceuticals, nutraceuticals, cosmeceuticals and medicines.

Along with bioproduct cultivation, growers will decarbonize the atmosphere, clean water and remediate degraded and destroyed ecosystems.

Food

Ecolanda growers will produce a wealth of conventional food, enough to feed five million consumers. Growers will cultivate microcrops such as algae-based foods, food ingredients and functional foods that deliver superior nutrition and health benefits. These foods increase resistance to disease vectors and fight disease when they attack the body. Consumers want to eat healthy whole foods close to the food's natural state.

Algae food ingredients have supplied stronger colour, flavor and texture to foods for millennia. Most ingredients today come from sea vegetables, macroalgae, because they have the longest history in foods.

Macroalgae, seaweeds, are highly visible and accessible. Our ancient ancestors commonly ate sea vegetables because they were nutritious and easily gathered along seashores and lakes.

Algae components are intensely integrated in modern food supply chains. An Arizona State University study performed a market basket test across seven grocery stores.[152] The research found that 72% of processed foods included one or more algae compounds.

Algae ingredients in modern food include:

Beer and diet sodas as a clarifier to remove haze-causing proteins. Frozen foods – pies, pastries fillings, yogurt and ice cream as an emulsifier.

Dairy – whipped toppings, milkshakes, skim milk, evaporated milk, chocolate milk, ice cream, cheeses, cottage cheese, infant formulas, custards and instant breakfasts. High protein drinks – healthier protein, vitamins, minerals and micronutrients.

Algae microcrop nutrient levels vs popular foods per kg

Nutrient	Microcrops have _ times higher than
Proteins - building blocks for bones, muscles, cartilage, skin, and blood. They help manufacture enzymes, hormones, and vitamins.	2x > than soy 3x > than beef, fish, pork 6x > than eggs
Iron carries oxygen in the blood. Many women in childbearing years have iron-deficiency anemia.	30x > than beef, fish, pork 65x > than spinach
B12 vitamin helps the body release energy, regulates the nervous system, aids in the formation of red blood cells, and help build tissues.	3 to 4x > than animal liver
Magnesium heart rhythm, build bones, maintains the immune system and normalizes blood pressure.	2x > than spinach 5x > than tomatoes
Calcium regulates nerve transmission, blood clotting, hormone secretion and muscle contraction.	10x > than milk
β-carotene, pro-vitamin A, boosts immune system, helps skin, eyes, protects against CHD and cancer.	5x > than carrots 40x > than spinach
Chlorophyll helps fight cancer, speeds wound healing, cleanses liver of toxins, improves skin, digestion and weight control.	30x > than spinach 20x > than wheatgrass
EPA and DHA, omega-3 fatty acids improve eyesight, brain function and protect from CHD.	1,000x > than any land plant 1,000x > than beef, poultry

Fruits – fruit juices, syrups, jams and jellies. Sauces, gravies and soy milk – thickeners and emulsifiers.

Pâtés and processed meat — substitute animal fat with low calorie alternatives.

Algae vitamins

Algae absorb a wealth of mineral elements that concentrate about one third of its dry biomass. The mineral macronutrients include sodium, calcium, magnesium, sulfur, potassium, chlorine, and phosphorus. Micronutrients include iodine, iron, zinc, boron, copper, selenium, molybdenum, fluoride, manganese and nickel. The biomass also includes high-quality vitamins, other minerals and trace elements.

One tablespoon of algae powder provides the same amount of calcium as ½ cup of milk, 1½-cups of soybeans, 11 carrots, or 22 tomatoes. A tablespoon provides the same amount of magnesium as 2½ cups of milk, ½-cup of soybeans 9 carrots or 5 tomatoes. A tablespoon gives the same amount of iron as 32 cups of milk, ⅓ cup of soybeans, 11 carrots, or 5 tomatoes.[153]

Algae provide a mineral profile superior to that of land plants, milk, eggs or soybeans.[154] Mineral availability from land plants, particularly legumes and grains, becomes compromised by phytic acid. Phytic acid binds the minerals, thus rendering them unavailable for absorption into the blood stream. In one investigation, phytic acid was undetectable in four species of algae. Iron absorption was 3.5 times higher for algae compared to rice.[155]

Algae iron is easily absorbed by the human body because its blue pigment, phycocyanin, forms soluble complexes with iron and other minerals during digestion. Phycocyanin makes algae iron more bioavailable. Unlike iron derived from terrestrial plants, the bioavailability of algae iron is comparable to that of heme iron in meats.[156]

Functional foods beneficially target functions in the body, beyond adequate nutritional effects. Common functional food enrichments are long chain omega-3 fatty acids-DHA/EPA, flavones, beta-carotene, lutein lycopene, fibre, proteins, catechins, beta-glucan and anthocyanins. Functional foods take all the forms of conventional foods including breads, soups, stews, pasta, beverages and snacks.

Major food retailers such as Amazon and Walmart are building supply chains that deliver food, medicines, pharmaceuticals and cosmeceuticals. Consumers will have the freedom to choose foods that prevent illnesses and other foods that treat specific diseases.

Market research shows that people strongly prefer following the advice of Hippocrates:

> *Let food be thy medicine and medicine be thy food.*

Algae support the global food system as useful ingredients and valuable compounds for functional foods. Consumers will choose algae foods for their extensive medical benefits as well as freedom from allergens and empty calories.

Taste

Taste represents the critical variable for new food adoption. Consumers adopt functions ONLY if the food meets taste expectation. Algae provide of all the essential nutrients, vitamins, minerals, and trace elements essential for health and vitality. Algae can satisfy any appetite with a broad spectrum of aromas, colors, tastes, and textures.

Algae's high glutamic acid content stimulate taste receptors, amplifying taste differentiation. This increases the desire to consume algae foods for good taste. Most functional food studies develop new practical uses from algae nutrient extracts that enhance food palatability. Kelp is added to beer to enhance the malty taste while other algae components improve liquid clarity.

Plant-based meats

Ecolanda will create plant-based meats, PBM, for every taste.  Each bite of PBM delivers 2.5 times more protein than beef. Unlike beef, these meats supply healthy oils, omega-3 EPA/DHA. PBM contain less fat, fewer calories and great taste.

PBM reduces health risks for obesity, diabetes, cancer and heart disease. They are healthier than tofu or texturized veggie protein which are often used in synthetic meats today.

These foods allow omnivores to diversify their diets. They contain no allergens, contaminants or pesticide residuals. PBM reduces consumer prices. They reduce social costs with zero consumption of water, cropland, fossil fuels, inorganic fertilisers or agri-poisons.

PBM cultivation can decarbonize the atmosphere, clean water and provide beautiful colour for animals like flamingos.

Several fast-growing algae species like *Haematococcus pluvialis* react to environmental stress by metabolizing valuable oils, such as astaxanthin. The cells spontaneously produce the oil to protect itself from too much sunlight. Astaxanthin can reduce free radicals and oxidative stress and help the human body maintain a healthy state.

Astaxanthin provides clean healthy oils to animal feed and gives the pinkish-red colour to salmon, shellfish.

3D printed food

Tiny algae fines make a perfect food construction material for 3D printers. These devices offer fast,

consistent and flexible formation of PBM and dairy products. They also make reliable and delicious algae-based pasta, pizza and pastries.

Human and animal health

Algae foods and ingredients optimize health. The proof lies in the addition of target compounds to functional and health foods. Many health promoting algae compounds are not available in conventional fossil foods grown in soil.

Algae Compounds Provide Value for Functional Foods			
Deficiencies	**Major Organs**	**Major Systems**	**Diseases**
Micronutrients	Brain	Cardiovascular	Blood pressure
Vitamins	Eyes	Digestive	Hyperlipdemia
Minerals	Blood	Endocrine	Bleeding gums
Trace elements	Heart	Immune	Infections
Antioxidants	Lungs	Respiratory	Inflammation
Hormones	Kidneys	Circulatory	Cancers
Dissorders	Skin, hair, nails	Urinary	Immune
Mood	Hypothalamus	Nervous	Viral infection
Anxiety	Pancreas	Muscular	Bacteria infection
Psychotic	Liver	Integumentary	Injuries, burns
Personality	Thyroid	Reproductive	Diarrhea
Sexual	Pituitary	Skeletal	Diabetes
Development	Nerves	Lymphatic	Obesity

Medicines

Micronutrients may be the most important compounds found in algae biomass. In many countries, including the US, over 85% of people suffer from micronutrient deficiencies.

Field studies show that a 3-week treatment with a tablespoon of algae a day brings nutrient concentrations to healthy levels.

A cluster of conditions causes metabolic syndrome — increased blood pressure, high blood sugar, excess body fat around the waist and abnormal cholesterol or triglyceride levels.

These tend to occur together, increasing risk of the common diseases. The incidence of the metabolic syndrome steadily increases worldwide. Algae-foods repair metabolic syndrome.

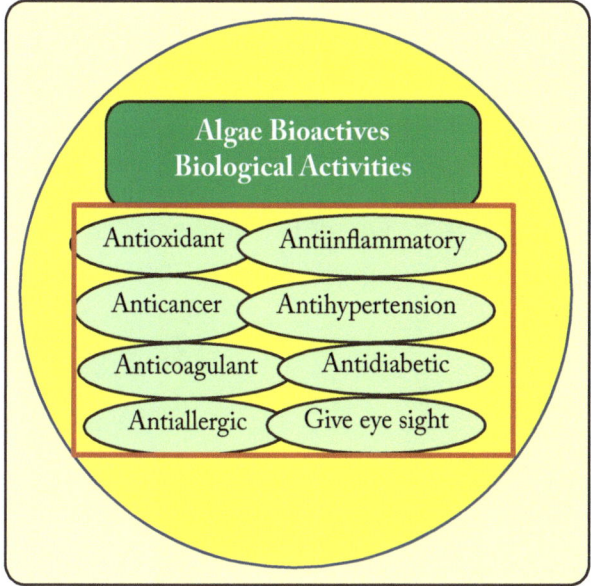

Several Asian countries where sea vegetables are commonly consumed exhibit very low metabolic syndrome rates. Several studies show that diets that include only 6 grams per day of algae improve critical inflammation biomarkers and blood lipids.[157]

Sodium alginate and other sea vegetable compounds are used in culinary physics at some of the best restaurants in the world.[158] Research shows sodium alginate pulls heavy metals including radioactive toxins from the body, such as iodine-131 and strontium-90.[159]

Bio-construction materials

Construction materials account over 8% of global GHG and has a much higher percentage of buried or burned wastes and hazardous materials.

Cement mining, mixing and applying emits about 4% of global GHG. Algae-based cement can stop those emissions. The carbon is sequestered as long as the material stays in place - several generations. At end-of-life, deconstructing cement structures, GHG emissions and wastes add substantial ecological costs to cement.

Asphalt may create more GHG than cement but is currently not included in climate models. The two-part construction process begins with the production of aggregates, asphalt and cement. The second step includes mixing, transportation, paving, compacting and curing the cement stabilized aggregate. At the end-of-life, most asphalt is deconstructed and buried.

Algae-based asphalt provides a green substitute for conventional petroleum asphalt. At the end-of-life for cement or asphalt, algae biomaterials can go through the BioRenew process to create new biomaterials or other bioproducts.

Construction-grade biofoam made from algae oil delivers 2x the tensile strength of concrete block and 3x the strength of wood construction. Biofoam is highly insulating and delivers several proofs: sound, odors, bugs, fire, earthquakes and wind. Insulation delivers an 80% lower cost of lifetime ownership saving utility costs. Biofoam cuts construction time by more than 50% and does not require skilled construction tradesmen.

Resin composites can replace steel beams with superior strength. Advanced biomaterials can replace materials that have a negative eco-footprint. Planes, boats, cars and trucks are being made with resin composites.

Food packaging such as pizza boxes commonly contain perfluoroalkyl substances, PFAS. These chemicals are linked to cancer, liver damage, decreased fertility, and increased risk of asthma and thyroid disease.[160] Algae-based biopackaging are clean and contain no dangerous chemicals.

Packaging creates massive waste. Biopackaging is strong, resilient and entirely recyclable with BioRenew. Bioplastics may be used for biodegradable packaging, windows, construction or many other applications. Bioedible packaging, bowls, chop sticks or knives, forks and spoons may be eaten or biocycled after use.

Bio-textiles

Textile production is the world's second most polluting industry after the oil industry. Total GHG emissions from textile production, 1.2 billion tonnes annually, represents 10% of global GHG. Textiles emit more than all international flights and maritime shipping combined.[161]

Textile mills generate one-fifth of the world's industrial water pollution.[162] Mills use 20,000 chemicals to make clothes, many of them carcinogenic. These toxic wastes typically run untreated into local waterways, damaging local residents' health. Textile factories produce over 5 billion tons of soot annually by burning coal.[163]

The primary raw material in the textile industry, cotton, consumes 20,000 liters of water to produce a cotton T-shirt and jeans. Cotton production uses 3% of annual global water consumption.[164] Industrial cotton production requires massive use of fertilisers, which contaminate local water bodies and cause dead zones. More pesticides are used in industrial cotton production than in any other crop.

Pesticides impose severe detrimental effects on young children. Children's immature livers and excretory systems may be unable to effectively remove pesticide metabolites. Research shows that low levels of pesticide exposure can affect young children's neurological and behavioral development. Evidence shows a link between pesticides and neonatal reflexes, psychomotor and mental development and attention-deficit hyperactivity disorder. Children exposed to pesticides suffer elevated rates of leukemia, brain cancer and soft tissue sarcoma.[165]

Algae contain plentiful fibres which can substitute for cotton, flax, bamboo and other textile sources. Algae textiles release zero GHG emission, clean wastewater during cultivation and use no pesticides.

Algae-based finishes and dyes for the textile industry present clean, non-allergenic, renewable and non-polluting alternatives to industrial methods. Algae dyes are typically brighter than industrial dyes.

Climate-controlled RAS aquaculture

Climate-controlled, land-based recirculating aquaculture systems provide unique advantages over ocean, estuary, lake or river farming. Land-based fish farms assure consistent, high specification water quality and temperature.

Climate-control allows for temperature and light variations that match the seasons, which is important for fish that naturally migrate.

SCAD filters biowastes from aquaculture systems and flows those wastes to hydroponic CEA where the nutrients are assimilated by vegetables.

Post hydroponic water still contains some waste nutrients and biosolids. This water flows to an algae biosystem for cleaning.

SCAD aquaculture uses state-of-the-art algae biofeed. Biofeed nutrition perfectly matches the fin or shellfish life stage. Sensors and monitors capture all the critical metrics, ensuring fish start healthy and stay healthy throughout the lives

SCAD creates specific recipes based on tested algorithms to optimize healthy aquaculture.

Feed and observation aquaculture platform

Save Biodiversity

Stressor management

Every animal grown faces stresses from a wide variety of sources. Most animal stressors are known and monitored. Field studies show that healthier nutrition significantly improves stress tolerance, which improves survivability and vitality.

Nrich enhances nutrition and ensures the animals receive the full set of macro and micronutrients, as well as vitamins, minerals and trace elements.

Algae-based biofeed are natural to fish, which improves both palatability and survivability. Algae are high in plant sugar, glucose, which makes the feed attractive. Superior nutrients are a waste if they were not bioavailable. Algae cells are so tiny, animals immediately assimilate the biofeed, which spurs growth at every life-stage.

High food bioavailability results in 10 to 20% less waste. Algae biofeeds are easier on animal stomachs than food grains. Efficient food uptake reduces producer costs because less food is needed. Less waste also lowers farmer waste management cost, which may be 20% of operational cost.

Many industrial producers dose their animals with antibiotics to reduce gastric distress from food grains and other foods unnatural to animal guts. SCAD has no need to use antibiotics because animals are able to quickly digest bioavailable biofeed.

Superior nutrition, superior animals

Nutritional and medical research reinforces the value of enhancing nutrition to improve health, stress-tolerance, disease avoidance, disease treatment and survivability. Microcrops offer superior nutritional attributes that are not available in land-based crops.

Succulents are plants that store excess water. No English word describes plants that store and deliver higher nutrient density, quality and diversity. **Nutralence** describes produce in a manner that offers relevant comparative metrics.

SCAD applies advanced metrics to monitor and assure high nutralence, which offers nutritional differences on five key dimensions.

Nutralence metrics

Nutralence includes these five elements:

1. **Nutrient quality**, the availability and stability of essential nutrients
2. **Nutrient density**, nutrients per bite
3. **Nutrient diversity**, the variety of nutrients
4. **Bioavailability**, the assimilation of nutrients easily and quickly without gastric distress
5. **Bioactive compounds**, over 300 bioactive compounds act to protect the consumer from disease or environmental stress

Nutralence metrics are shared with top nutritional institutions globally for validation. Studies have revealed significant positive nutralence differences compared with industrial agriculture crops.

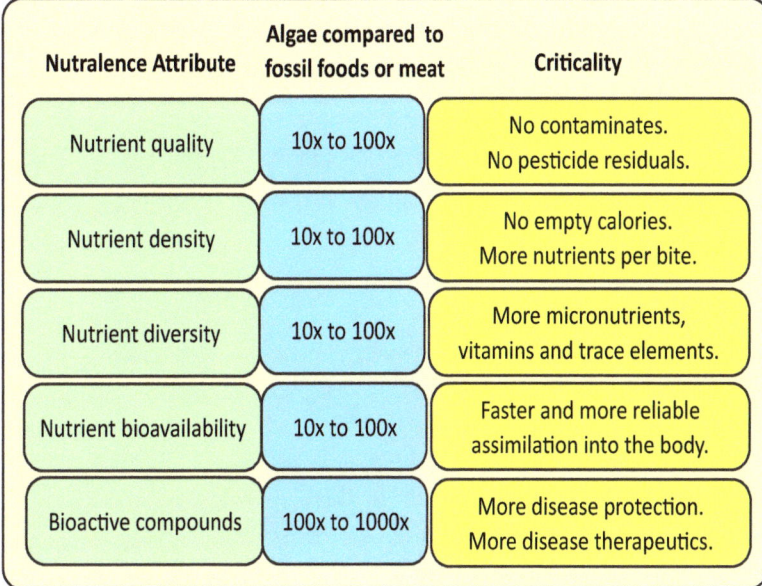

Nutralence Attribute	Algae compared to fossil foods or meat	Criticality
Nutrient quality	10x to 100x	No contaminates. No pesticide residuals.
Nutrient density	10x to 100x	No empty calories. More nutrients per bite.
Nutrient diversity	10x to 100x	More micronutrients, vitamins and trace elements.
Nutrient bioavailability	10x to 100x	Faster and more reliable assimilation into the body.
Bioactive compounds	100x to 1000x	More disease protection. More disease therapeutics.

Each nutralence attribute typically measures 20% or higher than modern industrial food and feed crops in SCAD cultivation. The table illustrates the nutralence advantage from experience with multiple crops.

Nutrient quality may be most difficult to measure. Microcrops and field crops fertilized by algae biofertiliser deliver produce that contains more than 10x more micronutrients and vitamins compared with industrial field crops. These metrics were assessed by an independent lab. Extensive research shows the value of higher nutrient quality, density and bioavailability.

Support a trip to Mars

Ecolanda will have a Learning Centre focused on how microcrops benefit society. Guests will have an opportunity to see how microcrops can support a trip to Mars and deep space.

The Ecolanda Learning Centre will apply the recommendations from the NASA 100-Year Starship project.[166] Algae and her sister microcrops can provide food, life-support, oxygen, waste biocycling, green chemicals, bioenergy, and building materials.

A 100-year Starship cannot possibly carry enough food for a trip to Mars or beyond. NASA calculates that six liters of algae water can produce 600 grams of food or 2500 calories. This is the average daily food requirement for an adult male. The six liters of algae water can simultaneously produce 600 liters of oxygen and consume 720 liters of CO_2.[167]

The carbon from the waste CO_2 gets transformed into hydrocarbons in healthy food for people, animals and plants.

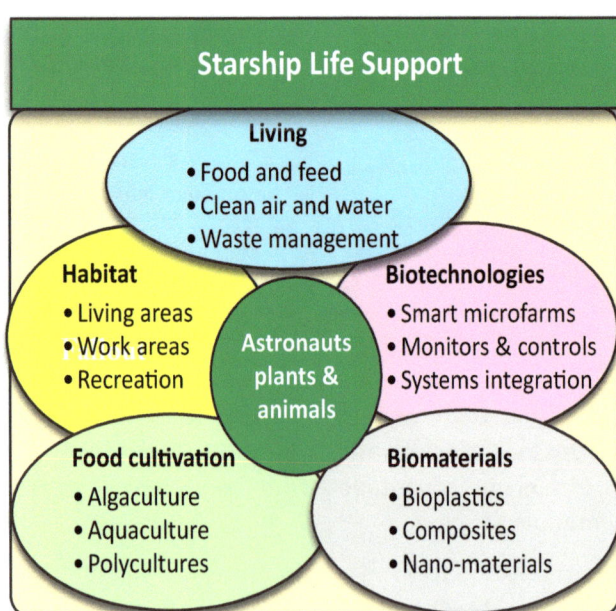

Human and animal waste products will need to be biocycled during the entire 100-year Starship mission. Gas waste, CO_2, methane, CH_4 and nitric and sulfur oxides are biocycled. Liquid wastes are biocycled into a host of clean bioproducts while algae clean the water for reuse. Human and animal solid wastes are biocycled into bioproducts for Starship life support.

The Ecolanda Learning Centre will highlight how microcrops can provide excellent medical support for the Starship crew. The crew are in top physical shape and want to maximize their health and vitality. Algae-based foods can be made into nearly any form, including ice cream, chocolate cake and plant-based steaks that imitate nearly any type of meat. Steaks are made with 3D printers that weave algae fibres into the meat to match the desired texture.

Algae-based functional foods, nutraceuticals, cosmeceuticals, pharmaceuticals and medicines sustain astronaut health. Functional foods help astronauts avoid illness while medicines fight disease threats when they penetrate the body.

Crew members will have an ongoing supply of recommended probiotics, antioxidants and phytochemicals.

The Starship cannot carry a pharmacy will all the necessary medicines. A novel Intelligent Algae Repository, (IAR) maintains a vast algae library of 600,000 algae cultivars that can be searched for novel compounds. Each cultivar contains about 1 million cells of the algae species that fit in 1 mm^2. Algae cultivation can grow the desired compounds quickly from the sample. The target compound can be extracted to resolve the malady. The IAR can produce a nearly infinite set of medicines just in time when they are needed.

17. Human and Environmental Health

Simplicity is the ultimate sophistication.
       – **Leonardo da Vinci**

Ecolanda's primary goal: *Lift the standard of living for everyone while improving the environment.*

Ecolanda listens and learns from each country's priorities for human and environmental health, economics and social. Project plans align with the country's goals to maximize benefits.

Stewardship

Human and environmental health are closely integrated. Polluted ecosystems, cities and towns impose a severe health toll on citizens.

SCAD makes commitments to remediate and restore healthy environments. The pace at which Nrich and BioRenew clean ecosystems can be tracked by National leaders. After 18 months, enough data exists to support National leaders to set new goals for carbon neutrality and clean environments.

Ecolanda ecological data will support other National goals such as water pollution and the avoidance of pollution vectors such as fossil fuels, chemical fertilizers and pesticides.

Blue water

Half of the world's hospital beds are filled with people suffering from a water-related disease.[168] Millions lack adequate access to blue drinking water. In developing countries, about 80% of illnesses are linked to poor water and sanitation conditions.[169] Healthy living requires safe, clean and affordable water.

Ecolanda not only saves water in agri-energy production but also creates extra blue water. The extra blue water will be shared with those most in need. SCAD will educate farmers in abundance growing methods that use Nrich biofertilizer that does not create toxic pollution in waterways. Abundance methods also minimize the need for pesticides and agri-poisons.

Malnutrition

One of every two people on earth, 3.8 billion, are food insecure, without reliable access to good food.[170] Over 9 million people die of hunger each year; more than the death toll for malaria, HIV/AIDs and tuberculosis combined.[171] Hungry people consume fewer than 2,000 calories a day.

Some communities must endure the severe pain and suffering associated with endemic malnutrition. Global hunger figures understate the much larger number of people who are malnourished due to the lack of quality food or suffer from micronutrient deficiencies.

Clinical signs and symptoms of micronutrient deficiencies include:[172]

- **Iron** – Fatigue, anemia, irregular heartbeat, dizziness and headaches.
- **Iodine** - Goiter, developmental delay, and mental retardation.
- **Vitamin D** - Poor growth, rickets, and hypocalcemia.
- **Vitamin A** - Night blindness, dry eyes, blindness and poor growth.
- **Folate** - Glossitis, anemia (megaloblastic), and neural tube defects.
- **Zinc** - Anemia, dwarfism, hepato-splenomegaly and hypogonadism.[173]

The high prevalence of hunger among women has led to malnutrition becoming the leading cause of death for children. About 3.1 million children die from hunger each year.[174] Poor nutrition accounts for roughly half the deaths for children under five.

SCAD Ecolanda plans to partner with National medical leaders for field trials to demonstrate how an 8-week supplement of 2 grams of algae a day can end micronutrient malnutrition. This natural biosolution for micronutrient deficiencies works for adults and for children.

Pregnant mothers with insufficient healthy food too often have premature or stillborn births. Some infants experience stunting or wasting due to lack of nutrients during womb life. These unfortunate infants are unlikely to ever experience a normal, productive life.

Protein-energy malnutrition, (PEM) creates a serious threat. If fetus fails to receive good nutrition during womb-life, major organs fail to grow properly. This results in stunting, slowing of linear growth and incomplete growth of major organs.[175] Behavioral changes occur such as irritability, apathy, decreased social responsiveness, anxiety, and attention deficits that lead to problems in school. Early childhood effects of micronutrient or PEM malnutrition may be irreversible.[176]

Freedom foods provide pregnant mothers and their fetuses with the nutrients they need for full-term, healthy births. These nutrient-packed foods ensure newborns avoid developmental disabilities so they can achieve their full potential.

Urban children may be taller than their rural colleagues due to nutritional differences. SCAD plans field demonstrations guided by National medical leaders to show how improving diets for

rural mothers and children eliminate stunting, wasting, premature births and developmental disabilities.

Obesity and diabetes

Many societies experience an epidemic of obesity and diabetes. Fossil foods are commonly made from low-nutrient food grains such as rice, wheat or maize. These hard, starchy and bitter grains are not palatable raw. Grains undergo extensive processing to make them palatable by adding extra fats, sugar and salt. Processed foods deliver empty calories – calories empty of nutrients.

Freedom foods provide tasty, high nutrient density nourishment. Algae are naturally soft with a sweet, neutral taste. Freedom foods are naturally low in calories, unhealthy fats, sugar and salt. Fossil foods drive the nosh response in the brain demanding more food. Freedom food provide a satiate brain signal that squashes nosh. It overrides nosh with a feeling of fullness.

Ecolanda creates goals for decarbonization, wastewater remediation and degraded cropland restoration. Each Ecolanda site creates goals for bioproducts in concert with the National Team.

Environmental health

SCAD make commitments in every Ecolanda project to improve environmental health. Similar to human health, needs vary by nation. Most areas have need for cleaning and restoring health to air, water and ecosystems. Each country makes a unique set of environmental goals.

The graphic provides examples of bioproducts sorted by value and volume.

High value, low volume

- Novel biochemistry molecules
- Sensors — medical & industrial
- Clean water, air and ecosystems
- Nutraceuticals and cosmeceuticals
- Biodegradable textiles and materials
- Organic LEDs, wiring and structures
- Malleable and insulated electronics
- Photovoltaics, recyclable electronics
- Battery membranes and electrolytes

Low value, high volume

- Emulsifiers, enzymes & preservatives
- Hygiene and absorbent bioproducts
- Construction materials and biofoam
- Aerospace structures and interiors
- Aerogels, coverings and insulators
- Biopaints, pigments and coatings

All SCAD Ecolanda bioproducts are:
- Biodegradable
- Net-zero waste and pollution
- Healthy for their intended use

Medium value

- Human food - functional and ingredients
- Plant-based meats and hybrid foods
- Superior nutrition, micronutrients, vitamins
- Specialty construction elements and fibres
- Automotive parts and furniture
- Packaging, films, coatings and bioplastics
- Paper, packaging, fillers and insulation
- Biofuels: oil, diesel, ethanol and jet fuel
- Biodegradable cementa and asphalta
- Feed - meat, dairy, farm animals & fish
- Ecological recovery — BioRestore
- Biotextiles and bright pigments
- Highly productive biofertilizer

Heavy metals

Many communities rely on deeper and deeper wells for drinking water. Deep wells increase the likelihood of contamination from deadly heavy metals such as lead, mercury, cadmium and arsenic.

Each of these toxins can cause severe damage to the developing brains, hearts and other major organs especially in children. The poisonous metals destroy adult health too.

Ecolanda biosystems can remove toxic heavy metals from water so they can do no harm. Heavy metal biomass may be made into bioplastics, bioresins and bioconstruction materials where they stay inert and non-threatening for decades.

BioRenew can remove toxic heavy metals from drinking water. What about children and adults that suffer from brain and central nervous system dysfunction as a result of ingesting heavy metals?

An algae health food, Spirulina, allows the body to assimilate the tiny algae cells that chelate with heavy metals. The cells can pass through the blood/brain barrier.

The body sluffs them off and passes them out of the body in the urine. In regions plagued with heavy metals, the ability of freedom foods to detox heavy metal poisoning can save millions from painful and ugly disability or death.[177]

Land reclamation

Large tracks of crop land have been abandoned due to erosion, salt, compaction and exhaustion. Expanding deserts, deforestation and invasive weeds threaten ruin of other lands. Rising oceans take a harsh toll on some of the most fertile crop lands, coastlines and river deltas.

SCAD creates plans and timelines to repair these lands and prepare them for restoration.

Coral reefs

Nearly half of the world's population live within 100 km, (60 miles) from a coastline.[178] In 1990, there were 10 megacities, with over 10 million people. The UN anticipates 41 megacities by 2030, with 36 on coastlines.[179]

Sea levels have risen an average of 40 cm, (1.3 ft), over the past 100 years. NOAA's models predict the rate of sea level rise to increase – substantially.[180] People living along coastlines are highly vulnerable to sea level rise.

An international study revealed that coral reefs protect billions of people along coastlines and river deltas from rising sea levels and damaging wave action.[181] *Nature Communications* calculated that **coral reefs reduce wave energy by 97%** and **wave height 84%**.[182] Reefs provide a critical naturally seawall protecting cities.

Coral reefs, rainforests of the sea, are home to the most biodiverse and productive ecosystems on earth. Corals are similar to fungi's structure in lichens on land. Corals and lichens depend on their symbionts, algae, for 90% of their nutrients and 100% of their pigments.

Coral reefs have a global economic value of over $375 billion a year. Reefs provide food and resources for more than 500 million people in over 100 countries. Unfortunately, coral reefs are dying, primarily as a result of industrial pollution. Nearly 90% of corals are under threat of extinction by 2030.

SCAD will work with global biotechnology experts to develop algae cultivars that can survive and thrive in warmer and more polluted oceans. These robust algae can be introduced to corals in order to improve coral survival.

Peace microfarm biosolutions

Peace microfarms preserve natural resources by biocycling waste streams, which may avoid conflict or war over land, water, fuel or fertilizer. A peace microfarm can clean wastewater and provide all the essential nutrients to avoid malnutrition for a community.[183]

Local microfarms can eliminate malnutrition and micronutrient deficiencies, in rural and urban areas. A single 50 m² microfarm can deliver enough Spirulina to cure **1,350 children** and/or pregnant mothers from the curse of premature birth due to malnutrition.[184]

Example: 50 m² (544 ft²) microfarm, (surface area), About 3m by 17m. Volume about 15,142 L (4,000 gallons) at 15 cm depth. Microfarm yields about 135 kg of Spirulina a year.

Antenna research shows that an 8-week Spirulina treatment with 100 g (total) resolves child malnutrition. Microfarms grow microcrops that produce healthy protein 30 to 50 times faster than field crops such as food grains. A microfarmer can harvest about 30% of the algae biomass daily or choose to harvest a higher percentage every two days during sunny weather. Growers can produce algae food and bioproducts year-round in many regions.

Each microfarm may employ several people. Microfarms will not make them rich but will provide the means for healthy food for their family and food to sell in their community. An estimated 100 small spirulina producers are growing food locally in French spirulina microfarms as far north as Normandy.[185]

Robert Henrikson tends an algae microfarm of his design. Robert created an excellent "Getting started checklist" for creative people interested in building a microfarm.[186]

Peace Microfarms: A green Algae Strategy to Prevent War explains how algae microfarms give growers the freedom to produce food, feed and other valuable bioproducts locally.[187] Peace microfarms avoid war by producing food and other forms of energy with minimal fossil resources. Resources saved eliminates the need to fight over scarce food production resources.

Microfarms can provide social justice, since only modest physical labor is required and there is no dust, pesticides, heavy machinery or poisons.

Women, physically handicapped and elderly people can grow highly nutritious food locally. Some microfarms have been designed for [Wounded Warriors](#) who have lost limbs in war.[188]

Families can support themselves and stay in the communities they love.

A microfarmer may cultivate one of many algae species. She may choose to dry the product or sell it fresh locally. Fresh or frozen algae may be eaten directly with no processing or cooking.

Peace microfarms can produce healthy food independent of altitude, latitude, climate or geography. Production systems can be sited on empty lots, rooftops, urban gardens, rail rights-of-way, and balconies. Microfarmers can grow enough food in cities to feed the entire city.

Microfarms will create good jobs for families that live in inner-cities, slums, food deserts and rural communities. Urban microfarms can significantly reduce transportation costs and black soot pollution from diesel truck engines because local food production nearly zeros out transport. Microfarms biocycle nutrients and grow excellent fertilizer and animal feed for urban farmers and family animals.

SCAD will create public-private partnerships to build microfarms and support growers.

Triple bottom line

Sustainable systems, also called triple bottom line, drive decisions across the Ecolanda verticals.

Social development includes far more than job creation to ensure people equal opportunity for jobs, training and projects. SCAD will have a strong community outreach process that builds local engagement and economic impact.

SCAD has created the architecture for Ecolanda to be the highest productivity agri-energy centre in the world. Economic development works closely aligned environmental stewardship.

Environmental emphasis begins with processes that ensure health and safety for all associates and the community. Sustainable production focuses on minimizing the use of extracted resources. Resource efficiency, like energy efficiency drives cultivation decisions.

SCAD not only plans to prevent pollution but also to use a suite of proprietary and public biological tools to restore degraded ecosystems. SCAD has a strong commitment to replenish lost biodiversity. Environmental restoration provides the foundation for biodiversity recovery.

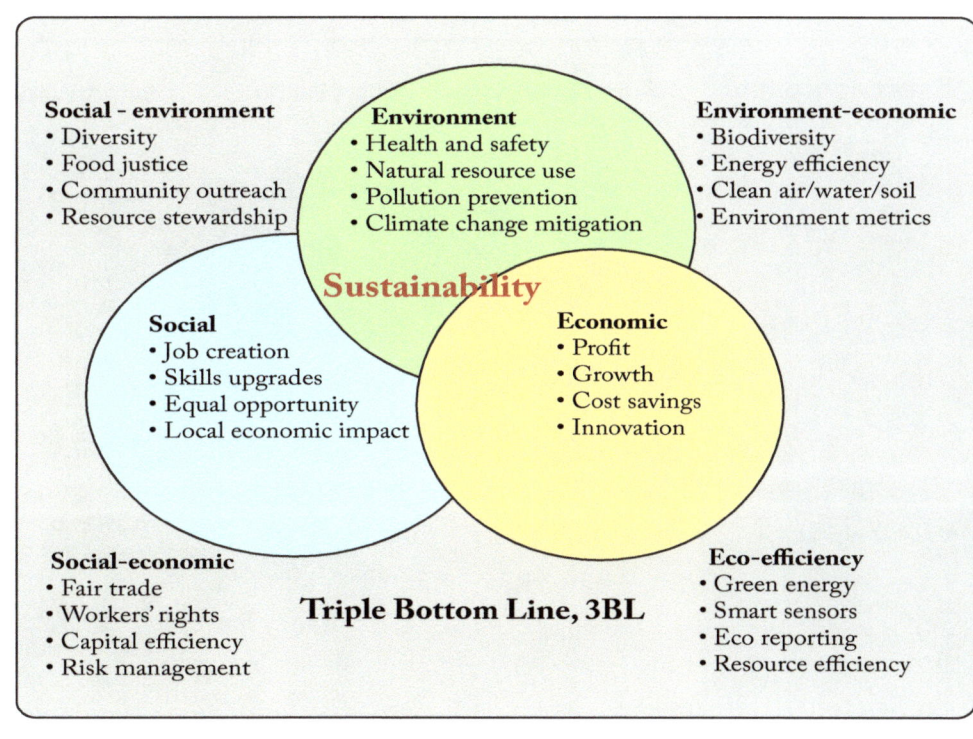

Summary

Ecolanda will provide a global benchmark for biosolutions that save biodiversity, save our children and save our environment. The metrics generated will serve policy makers so that they can understand the value for going beyond green to emerald systems.

Ecolanda will lift citizens' quality of life with 10,000 good jobs and over 50,000 support jobs. People will benefit from healthy food and affordable housing. Comprehensive education and training will continually improve living standards and economic development.

Economic growth will occur while producing healthier and more affordable food and emerald energy with zero waste and negative pollution. Ecolanda plans to cultivate enough food for 5 million people.

SCAD produces 90% of needed biofertiliser internally for all forms of food, feed and fibre cultivation. SCAD produces 80% of needed animal feed internally. Animal production. includes a million fish a year, plus other meat and dairy animals. The emerald bioeconomy, below, will demonstrate global leadership in addressing climate change by capturing and reusing over 70 million metrics tons of CO_2 annually.

Ecolanda will clean and restore millions of liters of blue water, which benefits the community. Nrich and BioRenew will remediate and restore thousands of hectares of degraded land and ecosystems. Restoration will allow the return of beautiful biodiversity.

What defines success?

Ecolanda succeeds when citizens and associates agree that together we have resolved four of the greatest challenges of our time. We will build:

1. Robust economic and social growth and development while cleaning and restoring our environment.
2. A 7-generation sustainable emerald bioeconomy.
3. Clean and restored ecosystems that naturally attract biodiversity repopulation.
4. Clean and healthy food supply chain and freedom foods that deliver superior nutrition and taste with no pollution or waste.

Let us act together and save biodiversity and save our children and their children.

Smart Agriculture	Smart Energy	Smart Water	Smart Waste	Smart Ecocity
Waste by type	**Total energy used**	**Fresh water used**	**Nutrients recovered**	**Energy efficiency**
• Waste recovered	• Total energy made	• Alternatives used	• GHG by type – used and lost	• Eco efficiencies
• Gas, liquid, biosolid	• Solar, wind, hydro	• Brine, waste, ocean, recycled	• Waste collection Waste efficiencies	• Resident sat
• Water efficiency	• Tides, geothermal	• Water efficiency	• Phosphorus, carbon, others	• Green transport
• Feed efficiencies	• Efficiency metrics	• Clean water		• Health, happiness
• Productivity	• Storage efficiency			• Sustainability

John G. O'Hare, Managing Director SCAD Queensland, Hong Kong, Malaysia & India

John plans, designs and leads the buildout and operations for all SCAD Ecolanda projects. He has responsibility for site selection, capital raising and building the SCAD leadership team.

Each SCAD Ecolanda™ benefits from John's 40 years of R&D in sustainable agri-energy systems.

He created the first architecture and plans for the world's finest circular biosystems that improve health for people, producers and our planet. These highly productive biosystems produce food, medicines, green energy, clean water and other valuable bioproducts

Ecolanda operates productively using minimal or no extraction of fossil resources while creating zero waste and zero pollution. These biosystems go beyond net-zero carbon and capture and reuse millions of metrics tons of carbon a year. The Ecolanda bioeconomy cleans and restores our environment.

John has used his site-selection matrix to evaluate over 50 potential Ecolanda sites over the past 20 years in China, Australia, Mexico, Greece, Malaysia, Ireland, Great Britain, Ghana and Andhra Pradesh, India.

John began his career as a Royal Australian Air Force Squadron during the Vietnam conflict. He followed that experience working for 7 years for Qantas Airways. He farmed and developed a series of nurseries across Australia and the Asia-Pacific region.

John worked over a decade as a registered equity derivatives trader in Australia and the Chicago Board of Trade. He holds a degree in Business from the University Southern Queensland. He earned several credentials for options trading in Australia and retail and institution certifications on the Chicago Board Options Exchange.

John invests considerable time investigating renewable energy methods that maximise carbon capture and conversion to Emerald Energy that is completely clean.

Mark R. Edwards, Director, Biotech and Ecology

Mark pursues 7-generation stewardship. This involves global R&D on emerald biotechnologies that drive sustainable bioeconomies for 140 years.

Ecolanda offers opportunities for stewardship with nations. We jointly design, build and operate abundance biosystems that biocycle carbon and other nutrients to eliminate the need for extracted fossil resources, waste or pollution.

Abundance methods avoid the need for war over scarce and increasingly expensive fossil natural resources. The BioRenew process biocycles waste streams and preserves precious resources for our future generations.

Mark graduated from the U.S. Naval Academy where he studied mechanical engineering, oceanography and meteorology. He earned an MBA and PhD in strategic marketing at ASU. He served as a director for the largest global food and transportation company and assisted in a series of successful food, agribusiness and technology start-up companies. He served as a professor at Arizona State University for 39 years, teaching food marketing, agribusiness, leadership, innovation, entrepreneurship and sustainability.

Mark founded and operated as CEO of a successful software and assessment firm, TEAMS Intl., for 22 years. He served as principal consultant and created advanced metrics for over 600 firms globally, including 7 of the 10 *Fortune Most Admired Companies*. He invented many innovative and award-winning metrics, including 360° Feedback. TEAMS Intl, the leader in 360° Feedback, won the prestigious Inc. 500 award for growth, leadership and profitability. TEAMS sold to an international consulting firm in 1999.

Mark authored both a business and a science bestseller. Seven books of his 21 books in the *Green Algae Strategy* series on sustainable and affordable food, water and energy won international best science and environment book awards. Mark has published over 150 business and scientific journal articles and writes the most popular blog in the algae industry, *Algae Secrets* for *Algae Planet*.

[1] https://www.oecd.org/environment/resources/biodiversity/G7-report-Biodiversity-Finance-and-the-Economic-and-Business-Case-for-Action.pdf
[2] https://www.sciencedirect.com/science/article/abs/pii/S0006320718313636
[3] https://eatforum.org/content/uploads/2019/01/EAT-Lancet_Commission_Summary_Report.pdf
[4] https://espresso.economist.com/e40d53ca19cd28f7dae77368fab8df4d
[5] https://pubmed.ncbi.nlm.nih.gov/30172774/
[6] Karn Vohra et al. Global mortality from outdoor fine particle pollution, *Environmental Research*, Vol 195, 2021, 110754.
[7] Edwards, Mark R. *Freedom Foods: Superior Nutrition and Taste without Pollution or Waste*, 2011.
[8] Edwards, Mark R. *Ecolanda: The First Emerald Circular Economy*, 2021.
[9] Edwards, Mark R. *Emerald Renaissance: World Hunger Solutions Healthier for People, Producers and our Planet*, 2019.
[10] https://www.worldwildlife.org/publications/living-planet-report-2020
[11] de Vos, Jurriaan et al. Estimating the normal background rate of species extinction, Conservation Biology, 29(2):452-462.
[12] Scott, J.M. 2008. Threats to Biological Diversity: Global, Continental, Local. U.S. Geological Survey, Idaho Cooperative Fish.
[13] https://royalsociety.org/science-events-and-lectures/2021/02/dasgupta-review/
[14] https://www.worldwildlife.org/publications/living-planet-report-2020
[15] https://ipbes.net/global-assessment
[16] https://www.cbd.int/conferences/post2020
[17] Edwards, Mark R. *Peace Microfarms: A Green Algae Strategy to prevent War*, 2014
[18] https://www.theguardian.com/environment/2021/mar/15/revealed-seafood-happening-on-a-vast-global-scale?utm_term=e68cafd3f6d479ab4ccf6c69a2d6cd0e&utm_campaign=GreenLight&utm_source=esp&utm_medium=Email&CMP=greenlight_email
[19] https://advances.sciencemag.org/content/6/9/eaaz3801
[20] https://finance.yahoo.com/news/mexico-vast-tree-planting-program-090024205.html
[21] https://waterfootprint.org/media/downloads/Hoekstra-Mekonnen-2012-WaterFootprint-of-Humanity.pdf
[22] Edwards, Mark R. *Abundance: Sustainable Fossil-free Foods with superior Nutrition and Taste*, 2010
[23] https://theconversation.com/fewer-crops-are-feeding-more-people-worldwide-and-thats-not-good-86105
[24] https://www.bioversityinternational.org/mainstreaming-agrobiodiversity/
[25] Edwards, Mark R. *Emerald Renaissance: Restore Health to People, Producers and Our Planet*, 2019.
[26] Edwards, Mark R. *Ana Feeds Our World by 2040: Miracles with Nature's Nano-cell Biofactory*, 2018
[27] Edwards, Mark R. *Ecolanda: The First Emerald Bioeconomy*, 2021.
[28] https://www.cnn.com/2003/TECH/science/05/14/coolsc.disappearingfish/
[29] Verhoeven JT, Agricultural use of wetlands: opportunities and limitations. *Ann Bot*. 2010;105(1):155-163.
[30] Miller GT (2004), *Sustaining the Earth*, 6th edition. Thompson Learning, Inc. Pacific Grove, California. Chapter 9, 211-216.
[31] https://www.ewg.org/foodnews/dirty-dozen.php
[32] https://www.ucsusa.org/resources/whats-driving-deforestation
[33] https://www.nationalgeographic.org/activity/save-the-plankton-breathe-freely/
[34] Pimentel, D., et al. Ecology of Increasing Diseases: Population Growth and Environmental Degradation. *Hum Ecol* 35, 653–668 (2007). https://doi.org/10.1007/s10745-007-9128-3
[35] https://phys.org/news/2020-06-heat-trapping-carbon-dioxide-air-high.html
[36] https://ourworldindata.org/food-ghg-emissions
[37] https://www.worldwildlife.org/threats/deforestation
[38] file:///Karn Vohra et al. Global mortality from outdoor fine particle pollution generated by fossil fuel combustion/ Results from GEOS-Chem, Environmental Research, Vol 195, 2021, 110754.
[39] Beaulieu, J.J., et al. Eutrophication will increase methane emissions from lakes and impoundments during the 21st century. *Nat Commun* **10,** 1375 (2019). https://doi.org/10.1038/s41467-019-09100-5
[40] WWAP (2017). The United Nations World Water Development Report 2017. Wastewater: The Untapped Resource. Paris: United Nations World Water Assessment Programme, UNESCO.
[41] https://academic.oup.com/humupd/article/23/6/646/4035689
[42] https://www.nature.com/articles/s41574-019-0273-8
[43] https://www.nimh.nih.gov/health/statistics/attention-deficit-hyperactivity-disorder-adhd.shtml
[44] https://adhd-institute.com/burden-of-adhd/impact-of-adhd/social-impact/
[45] https://drawdown.org/solutions/abandoned-farmland-restoration

[46] file:///Karn Vohra et al. Global mortality from outdoor fine particle pollution generated by fossil fuel combustion/ Results from GEOS-Chem, Environmental Research, Vol 195, 2021, 110754.
[47] Karn Vohra et al. Global mortality from outdoor fine particle pollution, *Environmental Research*, Vol 195, 2021, 110754.
[48] https://www-sciencedirect-com.ezproxy1.lib.asu.edu/science/article/pii/S0013935121000487#bib37
[49] https://www-sciencedirect-com.ezproxy1.lib.asu.edu/science/article/pii/S0013935121000487#bib55
[50] Magali Hurtado-Díaz, et al. Prenatal PM2.5 exposure and neurodevelopment at 2 years of age in a birth cohort from Mexico city, International Journal of Hygiene and Environmental Health, Volume 233, 2021, 113695.
[51] Bauer, S. E., Tsigaridis, K., and Miller, R. (2016), Significant atmospheric aerosol pollution caused by world food cultivation, *Geophys. Res. Lett.*, 43, 5394– 5400, doi:10.1002/2016GL068354.
[52] Siyuan Xiao, et al. Household mold, pesticide use, and childhood asthma: A nationwide study in the U.S., International Journal of Hygiene and Environmental Health, Volume 233, 2021, 113694,
[53] Global, regional, and national incidence, prevalence, and years lived with disability for 328 diseases and injuries for 195 countries, 1990–2016: a systematic analysis for the Global Burden of Disease Study 2016. Lancet 2017; 390: 1211–59.
[54] https://foodprint.org/issues/what-happens-to-animal-waste/?cid=906
[55] https://www.nrdc.org/issues/livestock-production
[56] Weselak, M.; Pesticide Exposures and Developmental Outcomes, J of Toxicology & Env **Health**: Part B. Jan2007, 10:1/2, 41-80.
[57] *Environmental Health: A Global Access Science* 2011;10(1):79-89. doi:10.1186/1476-069X-10-79
[58] http://www.panna.org/human-health-harms/children
[59] Vrijheid, Martine; et al. International Journal of Hygiene & Environmental **Health**. Jul2016, Vol. 219 Issue 4/5, p331-342.
[60] https://www.thelancet.com/commissions/pollution-and-health
[61] http://www.fao.org/3/CA0146EN/ca0146en.pdf
[62] https://www.wateraid.org/media/dirty-water-and-lack-of-safe-toilets-among-top-five-killers-of-women-worldwide
[63] http://www.unesco.org/new/en/natural-sciences/environment/water/wwap/wwdr/2017-wastewater-the-untapped-resource/
[64] https://www.who.int/news-room/fact-sheets/detail/drinking-water
[65] http://www.fao.org/land-water/news-archive/news-detail/en/c/1032702/
[66] https://www.ucsusa.org/food_and_agriculture/our-failing-food-system/industrial-agriculture/prescription-for-trouble.
[67] https://www.ecowatch.com/biodiversity-meat-wwf-2493305671.html
[68] https://pubs.acs.org/doi/10.1021/acs.est.0c05984
[69] Naidenko, O.V.; et al. Investigating Molecular Mechanisms of Immunotoxicity Added to Food. *Int. J. Environ. Res. Public Health* 2021, *18*, 3332. https://doi.org/10.3390/ijerph18073332
[70] https://pubs.acs.org/doi/10.1021/acs.est.0c05984
[71] https://earth.org/what-is-a-dead-zone/
[72] https://science.sciencemag.org/content/359/6371/eaam7240
[73] https://waterfootprint.org/media/downloads/Hoekstra-Mekonnen-2012-WaterFootprint-of-Humanity.pdf
[74] https://www.bakerinstitute.org/publications/EF-pub-BioFuelsWhitePaper-010510.pdf
[75] https://eatforum.org/eat-lancet-commission/
[76] Pimentel, D.; Satkiewicz, P. Malnutrition. In Encyclopedia of Sustainability, Volume Natural Resources and Sustainability; Berkshire Publishing Group: Great Barrington, MA, USA, 2013
[77] https://www.unicef.org/reports/state-of-worlds-children-2019
[78] World Health Organization (2014). WHA Global Nutrition Targets 2025: Stunting Policy Brief. Available online.
[79] http://www.fao.org/docrep/u8480e/U8480E07.htm
[80] https://www.thelancet.com/gbd
[81] http://articles.latimes.com/2003/jun/15/nation/na-diabetes15
[82] https://www.who.int/news-room/detail/11-10-2017-tenfold-increase-in-childhood-and-adolescent-obesity-in-four-decades-new-study-by-imperial-college-london-and-who
[83] https://www.cdc.gov/nchs/fastats/obesity-overweight.htm
[84] http://www.healthdata.org/news-release/vast-majority-american-adults-are-overweight-or-obese-and-weight-growing-problem-among
[85] https://www.cdc.gov/healthyweight/effects/index.html
[86] https://portal.nifa.usda.gov/web/crisprojectpages/0433776-food-markets--food-expenditures-and-marketing-costs.html
[87] http://uconnruddcenter.org/files/Pdfs/TargetedMarketingReport2019.pdf
[88] https://www.cdc.gov/chronicdisease/overview/index.htm
[89] www.med.umich.edu/1libr/Mhealthy/WhatAreEmptyCalories.pdf
[90] http://www.who.int/nutrition/topics/WHO_FAO_ICN2_videos_hiddenhunger/en/
[91] https://www.scientificamerican.com/article/soil-depletion-and-nutrition-loss/
[92] WHO. The global prevalence of anaemia in 2011. Geneva: World Health Organization; 2015. Available online.
[93] https://data.unicef.org/resources/state-of-the-worlds-children-2019/
[94] https://www.ers.usda.gov/topics/food-nutrition-assistance/food-security-in-the-us/key-statistics-graphics.aspx

Save Biodiversity

[95] file:///Boyle CA, Trends in the Prevalence of Developmental Disabilities in US, 1997–2008. Pediatrics. 2011%3B 27/ 1034-1042.
[96] http://www.cdc.gov/reproductivehealth/maternalinfanthealth/PretermBirth
[97] https://www.ncbi.nlm.nih.gov/pmc/articles/PMC4034518/
[98] https://www.advisory.com/en/daily-briefing/2019/12/02/middle-age-death
[99] https://pubmed.ncbi.nlm.nih.gov/22338036/
[100] Geist, H. J., (2002). Proximate Causes and Underlying Driving Forces of Tropical Deforestation. *BioScience*, *52*(2), 143-150.
[101] https://www.theguardian.com/environment/2020/jun/02/football-pitch-area-tropical-rainforest-lost
[102] http://www.panna.org/human-health-harms/children
[103] https://www.saveearth.info/deforestation/
[104] https://www.ucsusa.org/resources/whats-driving-deforestation
[105] Pearson, T.R.H., *et al.* Greenhouse gas emissions from tropical forest. *Carbon Balance Manage* **12**, 3 (2017). 2
[106] https://www.nationalgeographic.org/activity/save-the-plankton-breathe-freely/
[107] Jared Diamond, Collapse: How Societies Choose to Fail or Succeed, Penguin Books, 2011.
[108] Hongchang, Wang, 30 December 2009.
 web.archive.org/web/20091230071928/http:/www.library.utoronto.ca/pcs/state/chinaeco/forest.htm
[109] http://www.fao.org/state-of-forests/en/
[110] Wye Research and Education Centre. Riparian Forest Buffer Panel, 2002.
[111] https://ourworldindata.org/forests-and-deforestation
[112] The Emerald Forest ePrize in only partially funded by SCAD, Southern Cross Agri-Energy Development. We are looking for additional sponsors.
[113] We are looking for co-sponsors.
[114] https://www.dropbox.com/s/6vqrbb7wqoog7qf/Tiny%20Mighty%20Anna.pdf?dl=0
[115] http://news.nationalgeographic.com/news/2003/05/0515_030515_fishdecline.html
[116] http://www.fao.org/docrep/019/i3640e/i3640e.pdf
[117] https://www.ecowatch.com/one-third-of-commercial-fish-stocks-fished-at-unsustainable-levels-1910593830.html
[118] https://ourworldindata.org/rise-of-aquaculture
[119] https://www.seachoice.org/info-centre/aquaculture/wild-fish-in-feed/
[120] http://fishcount.org.uk/farmed-fish-welfare/numbers-of-fish-used-for-feed-in-aquaculture
[121] https://www.conserve-energy-future.com/GreenHouseEffect.php
[122] http://www3.weforum.org/docs/WEF_The_New_Plastics_Economy.pdf
[123] https://www.who.int/news-room/fact-sheets/detail/household-air-pollution-and-health
[124] http://www.friendsofgaviotas.org
[125] https://e360.yale.edu/features/could-abandoned-agricultural-lands-help-save-the-planet
[126] https://www.edf.org/sites/default/files/10333_Measuring_Carbon_Emissions_from_Tropical_Deforestation
[127] https://www.grain.org/article/entries/5976-emissions-impossible-how-big-meat-and-dairy-are-heating-planet
[128] Lal, R. Soil Erosion and Land Degradation In Soil degradation; Lal, R., Stewart, B.A., Eds.; Springer-Verlag: NY, 1990; 129–172.
[129] https://www.ag.ndsu.edu/publications/environment-natural-resources/environmental-impacts-of-brine-produced-water
[130] https://waterfootprint.org/media/downloads/Hoekstra-Mekonnen-2012-WaterFootprint-of-Humanity.pdf
[131] https://www.who.int/news-room/fact-sheets/detail/pesticide-residues-in-food
[132] Op cit.
[133] https://jbiolres.biomedcentral.com/articles/10.1186/2241-5793-21-6
[134] Loopstra, R, et al. Food insecurity and social protection in Europe: quasi-natural experiment of Europe's great recessions 2004–2012, *Preventive Medicine,* 2016, 89:44, 50.
[135] Graham, Linda and Lee Wilcox. Algae. New Jersey, Prentice Hall, 2008, 8.
[136] https://www.ers.usda.gov/topics/farm-practices-management/irrigation-water-use/
[137] https://water.usgs.gov/edu/activity-watercontent.php
[138] https://www.bakerinstitute.org/publications/EF-pub-BioFuelsWhitePaper-010510.pdf
[139] https://cfpub.epa.gov/si/si_public_record_report.cfm?Lab=NRMRL&dirEntryId=244350
[140] https://faculty.washington.edu/ktorii/stomata.html
[141] https://concretehelper.com/concrete-facts/
[142] https://www.thelancet.com/commissions/EAT
[143] http://connects.catalyst.harvard.edu/Profiles/display/Person/73533
[144] https://sci.waikato.ac.nz/farm/content/nutrientcycling.html
[145] https://www.amazon.com/Climate-Independent-Food-Survive-Freedom/dp/1479276847
[146] http://www.drugdevelopment-technology.com/projects/rimonabant/
[147] https://www.nature.com/articles/ijo2008235
[148] https://innoventondcts.mandela.ac.za/Microalgae-Technologies/Technologies-Coalgae
[149] https://www.ipbes.net/news/media-release-worsening-worldwide-land-degradation-now-'critical'-undermining-well-being-32

[150] https://www.history.com/news/6-civilizations-that-mysteriously-collapsed
[151] https://www.ncbi.nlm.nih.gov/pmc/articles/PMC4058318/
[152] Edwards, Mark R. *Green Algae Strategy*, 2008, 84.
[153] www.acam.org/blogpost/1092863/ACAM-Integrative-Medicine-Blog?tag=nutrition
[154] www.ncbi.nlm.nih.gov/pmc/articles/PMC5387034/
[155] https://www.ncbi.nlm.nih.gov/pubmed/12362796
[156] https://academic.oup.com/jn/article/137/12/2691/4670055
[157] apjcn.nhri.org.tw/server/apjcn/18/2/index.php
[158] https://www.triphobo.com/blog/best-molecular-gastronomy-restaurants
[159] https://www.ncbi.nlm.nih.gov/pmc/articles/PMC3654245/
[160] https://www.hsph.harvard.edu/news/hsph-in-the-news/pfas-health-risks-underestimated/
[161] https://www.downtoearth.org.in/news/environment/fashion-industry-may-use-quarter-of-world-s-carbon-budget-by-2050-61183
[162] https://www.nrdc.org/issues/encourage-textile-manufacturers-reduce-pollution
[163] Ibid.
[164] https://www.textiletoday.com.bd/water-pollution-due-textile-industry/
[165] Chen M, et al. 2015. *Residential Exposure to Pesticide During Childhood and Childhood Cancers: A Meta-Analysis*. Pediatrics.
[166] Edwards, Mark R. 2011 to 2013 DARPA and NASA selected research for papers and presentations in the Habitat and the Medicine tracks for the 100-Year Starship Symposium in Orlando and Houston.
[167] https://www.physicsforums.com/threads/algae-producing-oxygen-in-diving-cylinder.636257/
[168] https://thewaterproject.org/water-scarcity/water_stats
[169] https://www.dosomething.org/us/facts/11-facts-about-global-poverty
[170] http://www.who.int/nutrition/topics/ida/en/
[171] https://borgenproject.org/how-many-people-die-from-hunger-each-year/
[172] http://emedicine.medscape.com/article/985140-clinical#b4
[51] http://emedicine.medscape.com/article/912075-overview
[174] http://www.worldhunger.org/world-child-hunger-facts/
[175] https://www.sciencedirect.com/topics/medicine-and.../protein-energy-malnutrition
[176] https://www.ncbi.nlm.nih.gov/pmc/articles/PMC3137999/
[177] http://www.algaeindustrymagazine.com/can-algae-save-children-from-heavy-metals-poisoning/
[178] http://sedac.ciesin.columbia.edu/es/papers/Coastal_Zone_Pop_Method.pdf
[179] https://www.weforum.org/agenda/2016/11/the-10-fastest-growing-megacities-in-the-world/
[180] https://coast.noaa.gov/digitalcoast/tools/slr
[181] http://www.nature.com/ncomms/2014/140513/ncomms4794/full/ncomms4794.html
[182] https://www.nature.com/articles/ncomms4794
[183] Gershwin ME, Belay A (2007) *Spirulina in Human Nutrition and Health*. CRC Press.
[184] https://www.antenna.ch/fr/activites/nutrition/
[185] http://www.spirulinasource.com/spirulina/spirulina-farms/la-capitelle/
[186] http://www.smartmicrofarms.com/about/microfarm-business-opportunity/getting-started/
[187] http://www.amazon.com/Peace-Microfarms-Green-Strategy-prevent/dp/1480141208
[188] https://www.woundedwarriorproject.org/

www.ingramcontent.com/pod-product-compliance
Lightning Source LLC
Chambersburg PA
CBHW051913210526
45473CB00006B/1996